FOOD FEARS

To Walt, Jonathan and Christopher with love and gratitude

*"Ultimately, what sustainability requires of us is change in global society
as a whole…To start the global task to which we are called, we need a
specific place to begin, a specific place to stand, a specific place to initiate
the small, reformist changes that we can only hope may some day become
radically transformative.
We start with food". (Kloppenberg et al. 1996, 39)*

Food Fears
From Industrial to Sustainable Food Systems

ALISON BLAY-PALMER
Wilfrid Laurier University, Canada

ASHGATE

Published by
Ashgate Publishing Limited
Gower House
Croft Road
Aldershot
Hampshire GU11 3HR
England

Ashgate Publishing Company
Suite 420
101 Cherry Street
Burlington, VT 05401-4405
USA

Ashgate website: http://www.ashgate.com

British Library Cataloguing in Publication Data
Blay-Palmer, Alison
    Food fears : from industrial to sustainable food systems
    1. Food supply 2. Food industry and trade 3. Food
    consumption
    I. Title
    338.1'9

Library of Congress Cataloging-in-Publication Data
Blay-Palmer, Alison, 1961-
    Food fears : from industrial to sustainable food systems / by Alison Blay-Palmer.
    p. cm.
    Includes bibliographical references and index.
    ISBN 978-0-7546-7248-7
 1. Food supply. 2. Food industry and trade. 3. Food consumption. I. Title.

    HD9005.B63 2008
    338.1'9--dc22

                                                                    2007049102

ISBN 978 0 7546 7248 7

Printed and bound in Great Britain by TJ International Ltd, Padstow, Cornwall.

Contents

List of Tables

Acknowledgements

As with all books, this project drew on the energy and great work of many people. I am indebted to everyone who took part in the assorted research projects that informed this process. Food fears: From industrial to sustainable food systems began as a presentation to the Royal Geographic Society – IBG session on Alternative Food Systems in 2005. From there the idea for a book blossomed and got its legs. Betsy Donald was an outstanding and enthusiastic supporter and contributor to the project since the beginning. Her penetrating questions, lively writing example and insights have improved this book immeasurably. In particular, I am grateful for her contributions to Chapters 1, 6 and 7. Kevin Morgan provided unflinching mentor support and guidance. I cannot thank him enough. Visits with Pamela Courtenay-Hall and Gary Clausheide, both on and off farm, provided food for thought and much stimulating discussion about visions for the future of farming and food. Wayne Roberts and Lori Stahlbrand helped to crystallize and frame important final concepts. Thanks as well to Ashgate Publishing, especially Valerie Rose and her colleagues, for their unwavering support throughout the writing and editing process. Peter Blay, Irene Novaczek, Melanie Bedore, Sunny Lam, Mike Dwyer, Jennifer Miller, Michael Wolfe, Margaret Veinot, Alison Bell, Allister Veinot, Pauline Creedy, Margot Redmond, Jordan Kennie and Qun Jian together provided excellent background material and guidance.

This book is based on more than ten years of my own research on alternative food systems and community innovations. The Social Sciences and Humanities Research Council (SSHRC) of Canada funded the bulk of that research. The SSHRC MCRI project led by Meric Gertler and David Wolfe supported the research work in this book that informs Chapters 6 and 7. I am also thankful for the funding provided by Queen's University, Anne Prichard and colleagues at the Frontenac Community Futures Development Corporation, Rita Byvelds at OMAFRA as well as Craig Desjardins and his team at the Prince Edward, Lennox and Addington Communtiy Futures Development Corporation.

My family and friends also deserve my heartfelt gratitude for indulging and supporting me during the ups and downs of the thinking and writing process. Thanks particularly to Susan Blay and Jill Barton. I am also grateful to Derrick Blay, Faye Stevenson, Heather Amar, Barbara Turley-McIntyre, Valerie Bang-Jensen, Kevin Blay, Trevor Blay, Hannah Nelson-Teutsch, Elaine Field, Roan, Caitlin, Aidan, Ben, Victoria, MacLaren, Rebecca and Emma. To Walt, Christopher and Jonathan, I am especially thankful for your inspiration, patience and unflagging confidence. And finally to Walt, my deepest gratitude for your encouragement and editorial support when it was most needed – without you, there would not be a book.

The author and publisher wish to thank the following for permission to use copyrighted material:

Canadian Journal of Regional Sciences for permission to reprint excerpts from: Blay-Palmer, A. (2007) 'Who is minding the store? Innovation Strategy, the social good and agro-biotechnology research in Canada', *Canadian Journal of Regional Science*, 30(1): 39–56.

Elsevier Press for permission to reprint excerpts from: Blay-Palmer, A. and Donald, B. (2007), 'Manufacturing fear: The role of food processors and retailers in constructing alternative food geographies', in M. Kneafsey, L. Holloway and D. Maye (eds.) *Constructing 'Alternative' Food Geographies: Representation and Practice*, Elsevier Press (accepted).

The journal *Lien social et politiques* for permission to reprint excerpts from: Betsy Donald et Alison Blay-Palmer (2007), 'Manger biologique a l'ére de l'insécurité. Lien social et Politique- RIAC, 57', Printemps 2007, pages 63–73.

And, Statistics Canada for Table 4.1: 'Average Total family Income and Sources of Family Income for Farm Families on Unincorporated Farms, Canada', adapted from Statistics Canada website "Income of farm families, 2001 Census", http://www.statcan.ca/english/agcensus2001/first/socio/income.htm#4.

Chapter 1

Food Fear: Making Connections

Alison Blay-Palmer and Betsy Donald

Our mainstream food system is breaking down. Escalating rates of diabetes, cancer and obesity, excessive food miles, farm income crises, and growing food insecurity are just some of the problems identified with the current food system (Beliveau et al. 2006; Kirschenman 2005; Power 2005; Smith and Watkiss 2005; Appleby et al. 2003; Goodman and Watts 1997; Friedmann 1993; Goodman and Redclift 1991). Consumers are increasingly distanced from the physical, social and intellectual origins of their food by the cheap food system that privileges quantity and short-term efficiency over taste, sustenance, quality and the environment. Moreover, as will be argued throughout the book, the industrialization of food has created the conditions for food scares – including, for example, salmonella poisoning, Mad Cow disease, escalating rates of E coli, avian flu and most recently trans global ingredient scares that permeate the food system. But food scares, in conjunction with other social, environmental and diet-related health problems, have encouraged consumers to seek out 'alternative' food choices (Maye et al. 2007; Whatmore and Thorne 1997).

These alternative food choices are defined in many ways, with adjectives such as 'specialty', 'quality' and 'local' used to describe an array of food-supply network choices of specific ethnic, organic, fair-trade or artisan products. What these products seem to have in common is their appeal to quality-seeking consumers of food (Marsden and Smith 2005; Ilbery and Kneafsey 2000; Murdoch et al. 2000). Some of these quality seekers are searching for an 'authentic' product from their homeland whereas others are asking for food grown within a particular local foodshed (Morgan et al. 2006; Ilbery et al. 2005; Watts et al. 2005). Still others simply want products free from pesticides or herbicides, regardless of the source. The universal thread among these consumers, however, is that they are looking for something different from more mainstream agro-industrial producers or retailers. According to Whatmore et al., (2003, 389) not only do these alternative food networks construct a new trust between producer and consumer, but they also "redistribute value through the network against the logic of bulk commodity production". They go on to note that these alternative food networks are nourishing "new market, state and civic practices and visions". In this context, Europe is seen to be miles ahead of North American culture in terms of alternative food appreciation. According to this understanding, food 'alternativeness' has "come to be associated with an intensification of differences between (North) American and (Western) European food cultures and politics" (Whatmore et al. 2003, 389).

But even in North America factors such as Mad Cow disease, GMOs and public health concerns are pushing more and more people into eating foods that have one or more of the characteristics of being tasty, fresh, traceable, chemical free and locally produced or sourced. Although some of these consumers are motivated by the style or status that comes with consuming specialty food, others are motivated by deeper philosophical concerns. The alternative Italian movement 'Slow Food' has caught on in North America – a blend of politics, social consciousness, taste and sensuality. With 160 *convivia*[1] in the US and over 30 in Canada, Slow Food is seen as an alternative to a rapid-fire, fast-food, North American lifestyle (Slow Food Canada 2007; Slow Food USA 2007). The movement emphasizes saving regional foods and small producers, seeking to revive and celebrate organoleptic pleasures–something seen as missing from North American cultural life. But according to some scholars, North America's 'fear of food' and 'fear of pleasure' are deeply ingrained in North American culture, with some pointing to America's puritan roots as the reason behind the general distaste on this continent for the pleasures of eating and drinking (Levenstein 2003; Tuan 1979).

That our fear of food may in fact be deeply ingrained in North American culture strikes one as interesting – especially as one wonders whether our fear of food could be one reason for the recent adoption, or in some cases co-optation – of 'alternative' food practices by mainstream agro-industrial food players. It is important to acknowledge the need to avoid false dualisms (i.e. good/alternative versus bad/industrial food systems) (e.g. Morgan et al. 2006; Goodman 2004; Smithers et al. 2005) and to recognize the muddied areas that surface when one attempts to delineate one system from the other (e.g. Jackson et al. 2007). The increasing forays into organic and fair trade food by mainstream processors, retailers and food service industries blur the lines between alternative and mainstream, conventional food systems (Allen 2004; Beus and Dunlap 1990). It is clear that on the ground, there are hybrid combinations that obscure the boundaries between the two worlds of food (Morgan et al. 2006).

One explanation for the inability of the alternative food movement to move beyond the margins may be confusion around exactly what it stands for (cf Maye et al. 2007). For some, alternative food has come to mean any type of food that is labelled 'organic', 'local' 'quality' or 'fair trade' (Maye et al. 2007; Ilbery and Maye 2005). This can lead to market confusion as consumers seek out 'alternatives'. In other cases consumers are looking for a quick fix to alleviate their food fears and engage with alternative food on a superficial level. This may result in the cooption of the benefit by dominant agro-food players as consumers lack the information to understand the finer details related to their food purchases. For example, in the fall of 2005, with the support of TransFair USA and Oxfam America, McDonald's launched it's own fair trade coffee in the north eastern US (Chettero 2005). A year later, Wal-Mart introduced processed and fresh organic products, underscoring the contradictory and ambiguous nature of alternative foods (Gogoi 2006).

But while recognizing that there are hybrid combinations that blur the boundaries between the two worlds of food, we acknowledge the existence of two systems (Morgan et al. 2006). As well, we find merit in analytically separating out the two systems as there were and continue to be attempts to move towards more sustainable food provisioning systems and away from more industrial food production regimes. Broadly defined, the latter food system tends to engage with food production as if

[1] Slow Food groups that meet regularly to share and learn about food.

food were a commodity like cars or widgets. In the industrial food system (IFS), there has been an emphasis on quantity and large-scale production. Larger farms are favoured as a way to make more money, and monocultures are the dominant production strategy. Chemicals and technology are used as a first line solution when resolving production and other challenges. On the processing, distribution, retail and food services side, the predilection is towards vertical integration in a food system chain controlled by corporations, with ingredients being grown, processed and shipped around the world. Beus and Dunlop (1990) identify conventional agriculture with: centralization; capital, labour and technology dependence; competition; the domination of nature; increasing specialization and narrowing of production resources; and exploitation of resources that privileges short term over long term sustainability. This contrasts with what will be referred to as the alternative food system (AFS). Beus and Dunlop associate alternative agriculture with decentralization and more local production; independence and self-sufficiency; community; harmony with nature situating humans as part of and subject to nature; diversity; and, restraint of resource use with an eye on the long-term consequences of production. A shift in approach is also evident in the delivery and accessibility of food as farmer and consumer relationships are recast. The rise of farmers' markets, direct sell farm stands, community supported agriculture (CSAs) and local organic box delivery services are examples of new forms of interaction. Therefore 'alternative' becomes, in this context, and by contrast more representative of a re-framing of the *entire* food system to "articulate new forms of political association and market governance" (Whatmore et al, 2003, 390) rather than an ad hoc adoption. In an ideal world, these new forms aim to be more environmentally, nutritionally and socially sustainable than what now exists (Maxey 2007; Ilbery and Maye 2005; Raynolds 2004; Allen et al. 2004; Marsden 2003).

The contrasts between the two food systems are particularly interesting when one recognizes the frustration for those working in alternative food movements and the inability to move beyond the 'alternative' status in the direction of a more sustainable, comprehensive food system. This marginalization frustrates those who see the need for a more ecological food production, distribution and consumption approach in light of evidence that suggests a sustainable model can help dissipate many of the externalities of the existing dominant food system (Maxey 2007; Pretty et al. 2005; Allen 2004; Hinrichs 2003). Like Allen (2004), we believe that these constituent-examples of the alternative food movement will remain marginal until we address some of the long-term structural as well as social and cultural norms prevalent in our food system,

> Much work still remains to be done. Now that the ideas and priorities of alternative food movements have taken hold, it is time for the next – even more challenging – step. Alternative agro-food systems must acknowledge and address the deeper structural and cultural patterns that constrain the long-term resolution of social and environmental problems in the agro-food system. (Allen 2004, i)

In particular, it is argued in this book that until we understand what drives people in North America to eat as they do, little progress will be made in moving the alternative food movement forward. To progress, we need a more nuanced understanding of the structures and processes that link the regulatory and policy environment with food production, processing, marketing, retailing, consumption, the environment and food in the construction of food systems (Morgan et al. 2006; Dupuis and Goodman 2005; Guthman 2004a; Whatmore 2002; Harvey 2000). The goal of this book is to take a step in that direction by: 1) unpacking food systems from the perspective that North Americans must deal with the complex, distanced relationship with their food that has emerged over the last 150 years; 2) acknowledging that the complexity in the food system has created uncertainty and chaos, manifested through, for example the Mad Cow crisis and the potential avian flu pandemic; and 3) recognizing that this in turn has precipitated a fearful relationship between consumers and the industrial food system. By trying to understand the institutional and cultural origins of why this 'fear of food' may be deeply ingrained in North American culture, our relationship with food can be re-framed through a rethinking of a more sustainable food system.

Fear and Food

It is useful at this point to flesh out the term 'fear'. Yi Fu Tuan (1979) divides fear into the two 'strains' of alarm and anxiety,

> Alarm is triggered by an unobtrusive event in the environment, and an animal's instinctive response is to combat it or run. Anxiety on the other hand, is a diffuse sense of dread and presupposes an ability to anticipate. It commonly occurs when an animal is in a strange and disorienting milieu, separated from the supportive objects and figures of its home ground. Anxiety is a presentiment of danger when nothing in the immediate surroundings can be pinpointed as dangerous. The need for decisive action is checked by the lack of any specific, circumventable threat. (Tuan 1979, 5)

In the context of food then, fear can precipitate a 'flee' reaction – for example, the immediate decision to stop buying beef when BSE was first announced – or more subtle and pervasive effects – for example, a distrust of GE food while continuing to eat food that contains GEs. It is also worth considering Boudreau's analysis of fear and its role in the political-social sphere. Using Robin (2004) as a starting point, Boudreau explains that,

> ...fear is a political necessity to maintain a sense of unity and to generate innovative action. Yet, the object of fear is consequently depoliticized: what people are afraid of is not worthy of discussion, as long as fear enables unity. In this view, fear connects people together through their simultaneous perception of threat. (Boudreau 2007)

As Boudreau explains, politically motivated and socially pervasive fear can at once unite *and* desensitize people to a source of fear. As we shall see in later chapters, food fear can be framed through this conflicted construction as it is used both to rally people behind regulations and then more subtly to coalesce consumers into markets for processed food products.

As we investigate the links between food and fear, a spatial perspective helps point to variations or similarities in institutions and modes of regulation. This in turn provides a pathway out of the complexities created by the IFS. In offering insights into fear-based societal interactions, Lawson (2007) explains,

> ... A broader range of geographic research can move us beyond fear and toward constructive and hopeful interventions in our world...[Geographers can talk about] how our questions, our priorities and our resource allocations might shift if we start from positions of hope rather than positions of dread and anxiety. (Lawson 2007, 335) (cf Harvey 2000)

Considering Fear and Food: Fleeing Versus Wariness

Interest in the role that fear might play in the rise of the alternative food movement (especially the increased North American sales of foods labelled organic) emerged as an unexpected pattern of answers developed during interviews into innovation in the Toronto food and beverage industry. Reinforcing the inclination to explore this avenue was work by Kneafsey et al. 2004. In their research on the (re)connection of consumers and producers through 'alternative' food networks, they too remarked on food fear as a recurrent theme. They observed that,

> Whilst not claiming that anxiety is the *only* driver of food consumption decisions and practices, our research suggests that it is one of the key factors pushing the growth of 'alternative' food networks. We did not set out to ask consumers about their anxieties; rather anxieties emerged as an important feature of consumer responses to contemporary food provisioning and in many cases, were revealed to be a motive for participation in 'alternative' food networks. (Kneafsey et al. 2004, 1)

The comments about fear that emerged in the Toronto research project were an unexpected by-product of original research on the innovative dimensions of food in the North American urban economy (particularly the rise of ethnic, organic and fusion foods). From comments made during this research, consumer fear represented an important motivator for firms when developing alternative food products, processes or designs. Consumers raised the issue of fear as a motivating factor for their alternative food purchases. Fear was also an implicitly recurring theme in policy-reports on consumer-buying preferences that were reviewed.

Take organic food as an example. The most important reasons people consistently give for buying organic are food safety issues related to BSE, Genetically Modified Organisms (GMOs) and the use of pesticides (DECIMA 2004; Kortbech-Olesen 2004; Environics 2001). This is reflected in consumers' attempts to sort out what they are eating. In a 2004 survey, 80% of US eaters said they strongly support labelling food with GMO content because of concerns about the safety of those foods (Bostrom 2005), while UK consumers cited food safety as the primary reason for buying organic food (Rimal et al. 2005). Consumers are concerned about the safety of conventional food and want to know what they are putting in their bodies (Whatmore 2002; Goodman 1999).

While food fears have precipitated new market opportunities for selling alternative food, they have also laid the foundation for consumer confusion. This confusion arises, in part at least, from points where alternative and industrial food systems intersect and blur. On the one hand, consumer concerns about food safety have spawned interest in food 'quality', shorter food chains and direct buying-selling relationships (Whatmore et al. 2003; Murdoch et al. 2000). On the other hand, when the food system precipitates a food scare, it seems that conflicting reactions occur in some consumers' minds (Aubrun et al 2005, 32). First food scares may be seen by the consumer as a personal failure and, "interpreted as confirmation that individuals need to make smarter choices, and that individual foods should be avoided" (Aubrun et al. 2005, 32) (cf. Halkier 2004 for research on Denmark). Accordingly, consumers interpret food scare events as their personal responsibility and conclude they need to be making 'smarter choices'. Second, as Aubrun et al. (2005) explain, "'healthy food' and 'healthy eating' are understood in comfortable, little-picture terms – as opposed to having anything to do with systems of production, marketing or cultural patterns" (Aubrun et al. 2005, 32). These 'little picture' terms represent the consumer's deconstruction of food system threats into manageable bits. It seems that since consumers manage one crisis at a time, they fail to be engaged with a food system *qua* system. Instead consumers focus on coping with individual food crises and so do not have the objectivity or the resources to act as agents of change to move the *food system* in new directions. As a result, consumers deal with compartmentalized food scares and miss the chance to change overarching systemic problems.

But in addition to creating food fears, the IFS then taps into those specific fears (i.e. it capitalizes on each compartmentalized food crisis) and offers solutions within the conventional system. Fat-free food products are a good example of this circular process. Fat-free 'healthy' alternative products are created to address consumer fears about life expectancy and weight gain. As we know there are connections between weight and the consumption of foods that are high in fat and sugar – a brainchild of the processed food industry. So in response to a problem created by their own products, the solution that emerges in the IFS is to create other, highly processed food products. In this way, the food products and systems have gradually become more complex and difficult to control over time.

The response by IFS processors to consumer anxieties about their food is not surprising, of course. Indeed, making money out of exploiting people's fears is not new. As food historian Harvey Levenstein (2004, 6) has put it, "when you have a culture in which food is the object of fear and loathing as well as love, there are people who are going to discover innumerable creative and inventive ways of exploiting these fears". Thus this book makes the argument that to move away from a food culture based even partly on fear there needs to be a reorientation away from the current IFS to a more ecological, social and economically sustainable food system. A shift is required to a new food system that is less prone to the creation of unmanageable complexity. It is further argued that this change must be systemic.

In order to understand how this broken food system was created, it is useful to examine possible origins of our deeply ingrained 'fear of food'. First a political economy analysis of the dynamics between different forms of capital, social regulation and agriculture set the conditions for capitalist penetration in food and agriculture. Second, layered onto the economic consideration is an analysis of the socio-cultural dimensions of our anxious relationship with food.

The Political Economy of Food

In general terms, the framework for reviewing the origins of food fears in the North American food capitalist system has been inspired primarily by Marxist regulation theorists such as Michel Aglietta (1987), Robert Boyer (1990), Richard LeHeron (1993), Harriet Friedmann (1993) and especially the ecologically-spirited Alain Lipietz (1995). Like many critical political economists, these scholars seek to explain how capitalist social relations came to be reproduced across time and space while simultaneously being marked by contradictions that challenged their ongoing reproduction. Neoclassical models take the continuation of capitalism for granted. By contrast, regulationists start with an explicit rejection of market equilibrium as the central organizing force within capitalism, positing instead that social reproduction is the central imperative underlying capitalism. This reproduction of social relations is not smooth, but rather undergoes periods of crisis, during which conditions are such that it is challenging to achieve the reproduction of social relations. Alternatively, there are periods of stability, during which conditions are such that capital is able to accumulate in a relatively stabilized way.

It is suggested in this book that to some extent the capitalist agro-industrialized food system has been able to accumulate and concentrate for many decades. Chapters 2, 3 and 4 show how a number of public and private institutional forms, social practices and norms acted to regulate and stabilize the accumulation of industrial food capital. Like Aglietta (1987) we believe that these *structural forms* are neither automatic nor inevitable, but rather develop during particular periods in capitalist development. According to this view, the current agro-industrial food system is in disequilibrium and therefore we are also in a period of great creativity and experimentation with regard to the formation of new and innovative institutional forms, social practices and norms that may make the system more sustainable. We provide more evidence of those innovations in Chapter 8, but now turn some of the social assumptions that have helped to construct the current IFS in its roughly 200-year history. Particular attention is paid to the role of fear in the creation of this system.

The Human-nature Divide: The Foundation for Food Fears

The human-nature divide was a necessary condition for the accumulation of the existing form of food capitalism. People needed to become distanced consumers of food instead of proximate producers living in balance with nature for the current food system to have taken root and flourished.

Margaret Fitzsimmons (1989) offers insightful commentary on the nature-society divide in general in her seminal paper, 'The matter of nature'. Fitzsimmons traces the divisions between 'urban' and 'rural' lifestyles and the resulting distance of urban dwellers from nature. She links the urban-nature divide to a wider breech between society and nature that severed the social from the natural or material world of everyday life. This severance led to a distanced relationship with nature, our bodies, and the food that we ingest to nourish our 'selves' (Whatmore 2002; Goodman 1999; Kneen 1993). It has been argued that this division is entrenched in North American and some European cultures through many avenues including Christianity. As Cronon (1996) explains, our perceptions about and relationships to nature are socially constructed and contextualized. For example, the bible directs Noah and his sons – and all of their descendants and followers – to lord over nature. They are counselled after the flood to,

> Be fruitful, and multiply, and replenish the earth.
>
> And the fear of you and the dread of you shall be upon every beast of the earth, and upon every fowl of the air, upon all that moveth upon the earth, and upon all the fishes of the sea; into your hands are they delivered. (Genesis 9:1–9:2)

The application of the covenant between God and Noah that humankind had an obligation to dominate the world and its creatures is a seminal precept in the creation of the nature-society divide. The division precipitated an attitude of supremacy by some Christians over nature. Feelings of separation from nature were reinforced through food taboos as certain foods were deemed off limits for certain religious followers. For example, prohibitions against eating pork are founded in Old Testament requirements to eat cloven foot, ruminant animals (Farb and Armelagos 1980). The foundational principle for humans to dominate nature persisted through the Enlightenment and became entrenched with the rise of modern society (Eder 1996). The attitude of domination facilitated the development of an "efficient relationship to nature" (Eder 1996, 145) and was reinforced during the rise of modernism and positivism,

> Modern Europeans added two components to the Christian recovery project [returning the earth to its original Edenic state] – mechanistic science and laissez-faire capitalism – to create a grand master narrative of Enlightenment. Mechanistic science supplies the instrumental knowledge for reinventing the garden on earth. The Baconian-Cartesian-Newtonian project is premised on the power of technology to subdue and dominate nature, on the certainty of mathematical law, and on the unification of natural laws into a single framework of explanation...science and technology hastened the recovery project by inventing the tools and knowledge that could be used to dominate nature. (Merchant 1996, 136)

The groundbreaking work of contemporary social scientists David Harvey (1989), Yi Fu Tuan (1979) and Klaus Eder (1996) provide historical insights into several salient features with respect to fear and food, the separation of society from nature, and the culture of control of our food that has emerged as a central feature of the contemporary American landscape. This culture has deep historical roots in the positivist-science over nature philosophy that facilitates the ascendancy and facile adoption of technology in North America. Yi-Fu Tuan (1979) provides a historical-cultural context for fear and geography linking fear, landscapes and control as he asks,

What are the landscapes of fear? They are the almost infinite manifestations of the forces for chaos, natural and human…In a sense, every human construction – whether mental or material – is a component in a landscape of fear because it exists to contain chaos. (1979, 6)

Searching for the roots of fear in Anglo-Saxon culture, Tuan describes the preoccupation by Victorians with the need to eradicate sin and control animal impulses. Nature as dangerous and unpredictable had to be tamed. In this reading of social change, the Industrial Revolution and the rise of modernism was a time when society brought order to its surroundings as a way first to control risk and second to reduce fear. In the context of the Industrial Revolution, the machine emerges as both the means and symbol for the containment and reshaping of nature.

The 'urbanization' of the twentieth century was a process of realignment of the relation between humans and the material world of everyday life. The 'city space' became the dominant living space of the twentieth century – and possibly the definitive physical manifestation of human control over nature. The city emerged as the contested site of both control and threat. In the context of food, urbanists became increasingly distanced from their food sources, underscoring the pervasive view that nature is external and primordial to urbanization. The need to control nature meant that we increasingly distanced ourselves from food as an 'intimate' resource (Winson 1993; Kneen 1993). According to Gussow (2001), this desire for control emerges from our fear,

The notion that we'll get total control over nature, I think comes out of our fear. I think we [North Americans] are a very, very fearful people and so we want to be in control and in agriculture this has been just a pathology…I am a gardener – I can understand that. My introductory essay in my book called *Chicken Little, Tomato Sauce and Agriculture*, talks about walking out in the garden and seeing the peppers thriving in one bed and not in the other bed and saying 'wouldn't it be nice to know why. Wouldn't it be nice to get it all under control'. (Gussow 2001)

Thus it is argued in this book that in North America, the rise of the industrialized food system reflects people's separation from nature and their desire to control the natural environment. But, over the course of the rise of the industrial food system, control turns to chaos as the system becomes too large to manage. City-dwellers are physically distanced from their food as food travels the globe from producer to consumer making some people wary about their food. Food scares create crises of confidence and undermine trust in the food system. The combination of distanced food sources and consequent food scares results in a pervasive societal and individual fear of food.

As detailed in later chapters, the effect of the market economy as a catalyst in the creation of a fearful food culture opens up spaces where industry can profit from the separation of humans from their food. The supermarket as the site of food procurement for the average consumer embodies this separation,

> The supermarket was inevitable, the result of concentrated urbanization, the automobile, refrigeration, vacuum-packing, and other technological breakthroughs. An impersonal link in the food chain, the supermarket is an anonymous clinical place, where food is processed, packaged, shrink-wrapped, stickered, and shelved – alienated from its natural sources. It is the natural habitat of non-dairy creamers, Kool-Aid, Cheez Whiz and Tang. It is a business that calls lettuce not wrapped in plastic 'naked'. (Kingston 1994, 52)

In many ways one could argue that the modern-day supermarket and its vast quantities of highly processed non-perishable food products are the ultimate example of capitalist penetration into the field of agro-food.

Capitalism faced unique barriers in converting food into commodities that are not relevant in other sectors. The need for large tracts of quality land for production and the vagaries associated with producing a 'commodity' under varying weather and physical conditions are some of the challenges that are difficult to manage in dealing with this commodity (Kloppenburg 1988; Mann and Dickinson 1978). While these are just some of the obstacles that may explain the apparent historical inability of capital to penetrate agricultural production, capital has nonetheless met with a high degree of success in overcoming material barriers. Since the early 1900s, as documented in Chapter 4, farming has been converted to a process in which the farmer must purchase the bulk of resources used in production, such as seed, feed and fuel. As Kautsky expounded in 1902 when describing a book by Simons on the American farmer,

> He [Simons] shows that agriculture is not stationary, that the law of increasing control by great capital interests is also felt in this field, only in another form than in industries. The development in agriculture takes place in such a way that the various functions of agriculture are transferred one by one to great capitalist concerns by the help of modern technical improvements. In this way these functions cease to be agricultural and become industrial. (Kautsky 1902, 152)

Kautsky then goes on to refer to steam driven farm sowing, harvesting and irrigation equipment that had to be rented from the 'capitalist' so that though,

> The rest of agriculture which has not yet become industrialized exhibits few signs of vitality and becomes ever more dependent on the transportation companies and the great capitalist industries which alone render its products available for the consumer.... These are the means by which the property of the farmers in the tools of agricultural production are being more and more restricted and concentrated in the hands of capitalist exploiters. The small farmers are not displaced by mammoth farms, but they become more and more dependent on great capitalist concerns. The social condition of the farmer approaches ever more that of the sweating boss in industry. He is not yet a wage worker, but he ceases to be an independent producer. (Kautsky 1902, 153)

Advances in science and technology strengthened agricultural transformation. This is particularly evident in the case of plant breeding and seed production, where up until recently the seed's natural characteristics presented a barrier to its commodification

(Kloppenburg, 1988).[2] In an attempt to break the biological barrier, capital has pursued the scientific and technological approach through the development of GMO seeds and plants and the privatization of plant varieties through Plant Breeder Rights and patenting (Blay-Palmer 2007; Doern 1999). While plant breeding has been around for thousands of years, recent advances in genetics and molecular biology constitute a new technical form. The most profound of these advances was the development in the 1970s of the technique of splicing Deoxyribonucleic Acid (DNA) from one species to another. The technology allows life forms to be genetically engineered to produce specific products, or to include certain characteristics that are considered 'desirable' as with neutraceuticals[3] and other forms of 'pharming'. Moreover, by crossing species barriers, these new life forms combine genetic material in previously impossible ways, thus 'outdoing evolution' (Kloppenburg 1988, 3). The ability to 'outdo' evolution of course is not something that is to be pursued as an end in itself. Rather genetic engineering should be pursued as an undertaking with profound economic and social implications. What makes new techniques of genetic engineering so desirable from a business perspective is that the vast majority of GE products introduced to date were developed to sell a broader array of chemicals.

Several observers have referred to the applications of the new genetic technologies as the beginning of a 'biorevolution' (Eisen 1994; Busch et al. 1991). Rifkin (1983) made headlines in the early 1980s with his book 'Algeny' when he went as far as to refer to the dawning of a new 'biotechnical age'. While biotechnology has been thought to promise the 'greatest revolution in history', there is also profound concern that this 'revolution' is uncontrolled and under-regulated. In many cases, technology has outstripped political and social capacity to consider the implications of emergent technologies such as GMOs (McKibben 2003). As a result, society is faced with known risks such as genetic and environmental pollution, as well as a host of unknown (and arguably unknowable) risks (Whatmore 2002) – in effect, genetic chaos.

Recent food scares have only heightened that concern and also raised more theoretical questions about the role of food in the current era. Food itself is a force to be reckoned with as new pathogens and foodborne diseases wreak havoc with human health and international trade. While new theoretical advances in the political ecology of 'becoming' confirm the importance of food as an actor, powerful biological metaphors offer an expanded context for analysis and a new language to frame our questions (Braun 2007). This builds on debates in biology that continue to simmer about the role of evolution, competition, cooperation and a reinterpretation of Darwinism called symbiogenesis (Hird 2007; Margulis and Sagan 2002). These

---

2 Hybrid corn is an exception and laid the early foundation for companies to control seeds. Developed in the 1920s and made available commercially in the 1930s, hybrid corn seed was only productive with its particular traits for the first year of planting. Replanted seed was not reliable and so farmers had to buy seed every year from a company or research station to get the benefits (usually higher yield) of the hybrid variety. Hybrid corn only became available in Canada starting in the 1940s (Agricultural Research Services 2006; Reid 2003).

3 The word nutraceutical is attributed to Dr. Stephen DeFelice who defined the term in 1989 as, "any substance that is a food or a part of a food and provides medical or health benefits, including the prevention and treatment of disease" (Kalra 2003).

lines of thinking are all apt as they provide metaphors for social scientists as we seek words to describe emerging social challenges and the potential for societal change. For example, recent work by Harris Ali and Keil (2006) on global infectious disease surveillance systems, points to the porous nature of global boundaries so that politics and distance dissolve in the face of health crises such as SARS.

In the case of food fears, we are particularly interested in how pathogens interact with social, economic, political, and ecological structures and how we need to think about food as an active being. Food as an agent of change, while not likely possessing any intent beyond survival and propagation, nevertheless exists and 'becomes' beyond our control. This is particularly salient in the context of work by Fischler (1988 in Kneafsey et al. 2004) and others such as Whatmore (2002) and Goodman (1999) who point to the intimate act of ingesting food from the 'outside world' into the body as 'the basis of identity' (Fischler in Kneafsey et al. 2004, 2). In this sense then to lose control over our food is to lose a connection with a part of our selves. Clearly then the 'intimate commodity' can provoke fearful reactions as food production and processing are increasingly distanced from the consumer. The separation of humans from nature and their food helps to construct the way society perceives and manages risk.

The Social Construction of Risk

Increasingly risk in the 21st century is mediated through a panoply of social constructions. Although grounded in 'objective science' through measurements and monitoring, the interpretation of and the translation of risk is socially mediated through experts at every step of the assessment and reporting process (Beck 1992; Bhaskar 1979). Given the increasingly complex physical environment, the number of experts and opinions that emerge about any given source of risk are often confusing and divergent (Carolan 2006). In the case of food, it became obvious as this book evolved that public trust in the industrial food system was gradually eroded over the last 150 years through a combination of political, social and economic conditions that combined to rescale and reshape our access to and knowledge about our food. Since the mid-1800s, we have gradually traded personal knowledge about our food and quality for variety and convenience. The cost has been an increasingly unpredictable food system that has brought us sickness and death thanks to a range of food scare problems including tainted carrot juice and spinach, Mad Cows and E. coli. To borrow from Beck (1992), the increasing complexity of modern society has constructed simpler, rescaled, reflexive relationship between people and their food as they seek to be directly reconnected with what they eat and the people who grow it so they can trust their food to be safe again. In this way people seek to recapture some of the power they lost to the industrial food system. It may be that by recasting the food system that we can reconfigure a hopeful relationship with our food.

Book Outline

The purpose of this book, then, is to unpack some of the structural as well as socio-cultural forces that have shaped and are reshaping the food system and to explore the social construction of food fears. In particular, is an interest in the role of new regimes of regulation that emerged to support an IFS. It is posited that while the system was created to improve conditions for consumers, ultimately it sowed the seeds for the complexity of the industrial agro-food system. As the system has grown increasingly complicated over the decades, public safety has been compromised and food scares emerged in North America and other countries. It is ironic in some ways that the increasing complexity of the industrial food system and its associated regulatory regimes, precipitate unexpected new variables as elements of the very complex system start to interact in unforeseen ways. In simple terms, for example, social migration as part of industrialization moved people into cities and restructured social interaction. An effect of this migration from rural to urban was a distancing of people from their food. As food production grew in scale, increasing levels of food safety regulations have been needed to safeguard food. However, food regulation has not prevented many tragedies that now seem inherent to an industrial approach to producing food. In fact, it seems that the complexity has grown to the point where the system has become unmanageable. So instead of developing a holistic approach to food that integrates human health, environmental and social issues, by and large, North America has reacted to food system failures on an ad hoc basis. This has created disjointed policies and regulations, that to a greater or lesser degree have been shaped more with corporate rather than public interests in mind. As a result, instead of seeing food as a system that includes human, animal, ecological, economic and social well-being, a disjointed and chaotic system has emerged.

While it is clear that Canada and the United States are separate on many fronts, it is argued that in the context of food systems, many of the insights from the empirical work in Canada can be extended to a more general understanding of current North American food trends and practices, and vice versa. There are two reasons for making this assumption. First, as is demonstrated in more detail in the following chapters, the regulatory and economic boundaries between Canada and the US have become increasingly porous through trade and security harmonization initiatives (Gilbert 2005; Jenson 1989). Second, in the context of food, Canadians import over 58% of our food from the United States effectively accepting the health and safety regulations established in the US for American food producers and processors (Statistics Canada 2007). For example, in a discussion of the reality surrounding the flow of meat across the Canada/US borders in the context of the Mad Cow crisis,

> ...it must be recognized that the Canada/U.S. border is an arbitrary line that has little to no effect on the safety of the beef consumed in North America. In fact, the international team of scientific experts that examined the U.S. investigation of the Washington State BSE case concluded that the U.S. case cannot be dismissed as an 'imported' case, and that both the Alberta and the Washington State cases must be recognized as being indigenous to North America. North American countries must, then, harmonize not only BSE regulations, but even more. (Oliver and Fairbairn 2004, 20)

The incidence of contaminated spinach and carrot juice in the fall of 2006 point to the seamless nature of food flows between Canada and the US. In this case, the consumption of California produce led to illness and death in Canada. In addition the North American analysis is informed by insights from the UK. The progressive nature of the regulatory, retail and consumer environments in the UK provide an instructive perspective on alternative approaches to food system evolution.

The next chapters, combine a review of existing literature with insights from on-going research into the institutional context for AFS and IFS to explain the evolution of the most relevant forces that shaped the North American industrial food system and associated food fears. The remainder of the book is divided into two sections: the first section, Chapters 2 to 5, focuses on the history and evolution of food fears in North America. Chapter 2 we explores the industrialization of the food system starting in the mid 1880s. The connections between food adulteration, contamination and the neophyte food regulatory systems in the US and Canada are described as they provide part of the foundation for the industrialization of the North American food system. The conceptualization of food as fuel in North America is also explored. This is undertaken in part through an examination of personalities from the early days of food regulation and processing in North America and includes well-known figures such as Olin Atwater and the Kellogg family. In this chapter it is argued that food fears are rooted in: the division created between society and nature; the industrialization and commercialization of food; the regulation and scientific domination of the food systems; and, the application of high technology to food production. As society evolved to be increasingly urbanized and 'progressive', food becomes increasingly packaged and convenient. There is also a survey of the emergence of science and technology and related research and development policies in the US, UK and Canada as critical factors in the acceptance by consumers of technological fixes to industrial food problems. Recent food scares emerge as an almost inevitable culmination of food industrialization and the vast distances that exist between farmers and consumers. With the institutional and regulatory context from Chapter 2, the stage is set for the following two chapters that explore the consequences of this industrialized food system. Chapter 3 discusses the commercialization and marketing of food and the dynamics that emerge as food is processed and branded. Much of the success of American food advertising and production in the 20th century can be traced back to a deeper historical separation between society and nature as described earlier in this chapter as well as in Chapter 2. Chapter 3 examines the links between: 1) increased food processing and the distancing of consumers from their food; and, 2) the rise of product branding and marketing to reassure consumers about food reliability and safety. The advent of the Kellogg brand name is traced as a case study of the strategies used to build consumer confidence in processed food. The book then turns to food retail and branding to examine M&S in the UK and President's Choice in Canada. This chapter covers the period from the early 1900s through WWII up to the early 1980s. The chapter ends with a discussion of the effects of advertising, particularly as it relates to children and the consumption of food high in sugars and fats. Chapter 4 explores the links between the huge economic pressures exerted on farmers, shifting production practices, ecological compromises and consumer food fears. Topics discussed in this chapter include the controversy surrounding genetically modified organisms (GMOs) and the mistrust they create for consumers; farm income instability as a force driving farmers to scale up their operations or to seek income off farm; and, factory farming and the related E. coli and water management issues that emerge from industrial scale farming. While Chapters 2 and 3 are grounded theoretically in the political economy and regulation theory literatures to explore the role of capital accumulation in the development of the IFS, Chapter 4 uses political ecology to explicitly engage with environmental concerns and sustainability issues. In Chapter 4 the power of different farm groups is unpacked as they operate within and beyond the agro-food industry and the various ways this shapes the way local ecologies can be considered by farmers.

Chapter 5 assumes a different tone as it explores the role of food as an agent of change in the industrial food system. This chapter combines the insights from the previous chapters and the gaps that are identified with respect to transparency and the protection of the public good. Chapter 5 we examines the relationship between food and disease. Alar, salmonella, BSE/ Mad Cow and Creutzfeldt Jakob Disease, and the threatened avian flu pandemic provide examples of the failure of the industrial food system. Chapter 5 describes the impact of food as an actor in the context of food scares. This chapter leads to the second part of the book that explores 'alternative' expressions that have emerged from the industrial food system.

Chapter 6 examines the role of consumers. The self-regulatory turn on the part of eaters and the rise of AFSs is examined with a focus on organics in the current climate of heightened security concerns since 9-11. Insights from web surveys and interviews are used to unpack how food fears drive consumers to self-regulate their food system in the context of the failed welfare state. Consumers are (re)turning to trust based relationships as one solution in their quest for safe food. In Chapter 7 the focus is on issues of food safety and how they mould the alternative food processing and distribution. Building from the work of Guthman (2004a, b) it is argued that by reacting to specific food safety issues created by the industrialization of food processing, the 'alternative' plays into the hands of the industrial food system. Empirical work confirms that the expression of 'alternative' in specific, codified terms enables the cooption of alternative market segments by dominant agro-food players. As a result, the AFS is repeatedly undermined as industrial food cannibalizes bits of the alternative and absorbs them as branches of the mainstream food system.

In the first seven chapters by pointing to: 1) the unmanageable complexity of the IFS as a threat to human, ecological and community health; and, 2) the lack of coherence in the AFS, an outline emerges of some primary contributing factors to the current 'broken' food system. In Chapter 8, the book concludes with a survey of food innovations that point to hopeful initiatives that have emerged as communities seek solutions to IFS challenges. Eating more from local food systems gives people hope that through the quotidian act of eating they can recapture some of their lost power. The book argues that to create a new food system calls for a reconsideration of relationships based not on competition, but that draws from cooperative and mutual interactions. Case studies are presented from Canada, the US and the EU to illustrate how the new food movement is gaining currency. Programs that currently escape the attention of regulators are presented as examples of ways that people are reconfiguring food systems from the ground up. The book concludes by proposing a new food system that offers a tentative template for a civic based, cooperative model respectful of ecology, society and the economy. Through hope there is the potential for lasting change founded upon a vision of a diversified, robust, renewed food system.

Chapter 2

The Industrial Revolution of Food

According to USDA's Economic Research Service, hazards in food cause an estimated 76 million illnesses, 325,000 hospitalizations, and 5,000 deaths in the United States each year. (Heller 2007)

Chapter 2 leads us through a survey of high points and trends in the evolution of the industrial food industry in the US, UK and Canada from the mid-1800s to the present. What emerges is the story of an increasingly complex food regulatory framework originally established to mitigate fear that now seems to be compounding consumer anxieties. Part of the story, traces the regulation of food in response to discoveries about the links between pathogens, food and illnesses. Connections are made between food adulteration, contamination and the neophyte food regulatory systems in the US, UK and Canada, explaining how they underpin the industrialization of the North American food system. Chapter 2 discusses the solidified authority of science that emerged following WWII and continued to consolidate itself into the 1980s. We explore the parallel move by governments towards support for more applied research and the increased importance of biological sciences in the food system that culminated in biotechnology. This chapter tells the story of public policies so tied to private interests that by the 1990s that the public good was threatened – the BSE crisis in all of the countries is a shared example of the almost inevitable result of food industrialization and the disregard for food as an embodied, personal source of nourishment and pleasure (Whatmore 2002; Goodman 1999). Theoretically this chapter draws on political economy and regulation theory to describe the gradual accumulation and concentration of material, human, social and political capital over time. By the end of the chapter, it becomes clear that: 1) the separation of government into disconnected, conflicting silos was established by the late 19th century and that this persists to the present, particularly in Canada and the US; 2) this results in a chaotic and cumbersome set of regulatory institutions that undermines our ability to effectively safeguard our food well-being.

The Industrialization of Food

The North American Food system has witnessed steady industrialization since the early 1800s. Once the 'intimate commodity', food gradually transformed into the 'industrial commodity' (Winson 1993) as we moved from a largely localized consumption practice to an industrial commodity system of mass consumption subsumed into the market economy (Murdoch and Miele 1999; Goodman and Watts 1997). As the industrial food system evolved, direct production-consumption connections weakened and eaters were increasingly separated from farmers and their food. The separation resulted from a physical distancing that created an emotional and intellectual divide between people and their food. However, the evolution of this relationship has not been a linear one extending from a past of total confidence in our food to a current relationship characterized by fear and uncertainty. In fact, the role of actors and institutions at every point along the food chain has shifted over time (Morgan et al. 2006; Walker 2004; Draper and Green 2002; Watts and Goodman 1997; Wright 1994; Winson 1993; Friedmann 1993). Part of this transformation occurred as increasingly sophisticated food processing technologies emerged at the beginning of the 19th century. It is also tied up with our scientific understanding of food.

Food Innovations: Food Preservation and Disease

It is useful to understand that the history described in this section sets the tone and intent for food safety and preservation goals in the many decades that follow. The emergent industrialization of the food process leads to an iteration between increased urbanization, growing distances between people and their food, and the scaling up of what will become the food industry. It emerges that as the industry grows, so do the problems. As a solution, the abiding faith in technology leads to more growth and consolidation.

The relevant pieces of history for the purposes of this book started to unfold about two hundred years ago when large strides were made in our capacity to store and ship food safely. Beginning in the early 1800s the ability to preserve food was revolutionized by new canning techniques. The first innovation of note came in 1809 when Nicolas Appert won the 12,000 francs prize from Napoleon for inventing a way to preserve food by using bottles to 'can' food. By heating food in bottles the French were able to supply their troops fighting overseas with 'indigenous' food (Scott and Elliott 2006, 77). Although various methods of food preservation had been used for millennia (for example, salting, smoking, fermenting and drying), the invention of canning allowed people to store food for a long period of time and permitted its transportation over long distances. The process was used to store meat, milk, vegetables and fruits. In 1819, William Underwood used tin 'canisters' (the root of 'can') in the United States to preserve food. The new technology facilitated voyages by the British navy and army for exploratory ventures in the Arctic and supported the spread of colonialism. Canning in iron and tin containers was such an important innovation its adoption was widespread by the mid 1800s. By the 1870s, the canning process was mechanized in keeping with other mass production processes of the time (Jackson 1979 in Scott and Elliott 2006). Improvements and innovations in refrigeration condensers and evaporators led to improved cold storage methods so that by the 1870s cold stored mutton and beef were shipped from Australia to England (Lund 2000).

At the same time as preservation methods improved, scientists developed a better understanding of germs, bacteria and their connection to food and health. In 1862 Louis Pasteur completed the first pasteurization test. From that point forward, pasteurization was used to kill unwanted bacteria, to make milk safer, and to extend food shelf life.[1] In the 1880s connections were established between infective organisms and food poisoning (Draper and Green 2002). From then on, science and technology were applied enthusiastically to food processing and handling to make it as safe and efficient as possible. The convergence of a growing awareness of microbiology and the shipping of food from the US to the UK led to one of the first recorded trans-Atlantic food poisoning incidents in the 1880s. On this occasion, undercooked hams infected with a parasite were fed to attendees at a public auction in Nottinghamshire. Over seventy cases of illness and four deaths were attributed to this first documented case of what came to be labelled food poisoning (Hardy 1999). One could argue that this signals a turning point in public and government attitudes about food safety and regulation and marks a change from private to public responsibility. It also acts as a landmark for the emergence of a global food system. The increased understanding of the links between food and disease led to calls for governments to protect the public through legislation. Public reaction in the US and the UK and the demand for regulation laid the groundwork for contemporary international norms for the inspection and control of food.

Government Regulation and Food in the UK

As the links were made between disease, bacteria and food at the end of the 19th century, there was a genuine sense of urgency to impose scientific assessment and government regulation onto the food preparation process. The creation of regulations took place as new challenges were being faced about: the implications of new technologies; new international relationships between countries and their existing or former colonies; and, emerging international trade regimes.

All of this occurred in the context of the fact that as early as 1753, the British physician Lind established the connection between scurvy and eating citrus (Drummond and Wilbraham 1958 in Lang and Heasman 2004, 104). This connection allowed people to prevent disease through better nutrition. By the early 1800s an additional concern started to become prominent as Germany and the UK focused their attention on food adulteration as evidence of intentional food contamination came to light. The German chemist Accum exposed the addition of cheap additives to food as unsavoury food processing practices. For example, alum was used to whiten bread and lead and copper salts were included in confections for colouring (Draper and Green 2002; see also Burnett 1989). Even in these early days of food processing on a larger scale, processors adulterated food in order to boost profits. In Britain, this prompted the formulation of the Adulteration Acts of 1860, 1872 and 1875 as attempts were made to shield the public from unscrupulous food processors and retailers. These Acts emphasized the composition of different foods such as milk and other basic food items, and focused on food safety.

The rapid urbanization of Britain along with the movement of people and germs around the globe as part of the spread of rail and sea travel in the mid-1800s fuelled

---

1 It is interesting to note that unpasteurized milk is one way that consumers in the present food system express their autonomy by opting out of the conventional system for a more trust-based farmer-eater relationship (Enticott 2003). This point is tackled in Chapter 7.

the dissemination of food related infection. At the same time, as Hardy (1999, 297) points out, street vendors of the mid-1800s sold, among other food products, spoiled fish disguised in fish and chips. These entrepreneurs – perhaps among the first fast-food vendors – provoked more regulation, this time to prevent unnecessary food borne illnesses and deaths from negligent handling and hazardous food content (Draper and Green 2002). But while food vendors were being regulated, there was reluctance in Britain to impose 'onerous' regulations on other parts of the food chain such as abattoirs. So, while there were examples of regulation to address specific high profile problems, overall, there was a preference to avoid over-regulating independent business (Hardy 1999).

Beyond the government's reluctance to act, UK consumers also served to dampen more enthusiastic food regulations. For example, attempts in the 1930s to protect food from flies by placing it under glass were unpopular as shoppers wanted ready access to be able to inspect the food they were buying. In the early 1900s, food-handling conditions in shops by UK food purveyors spread disease as hygienic standards did not include hand washing and what would be considered basic health standards today. Hardy reports, "glasses for example, were commonly only dipped in dirty water before being reused; personal cleanliness was uncommon among food-handlers, who regularly licked their fingers when dealing with wrapping paper, and blew into paper bags to open them. Hand washing after opening the bowel was generally neglected, and the habit of putting fingers into the nose and mouth while serving food could regularly be observed" (1999, 306). The lack of sinks or wash basins in WCs pointed to the challenges faced by public health officials in getting food vendors to use more sanitary approaches to food handling and hygiene.

At the same time, the regulation of food and health standards that did exist were very disjointed. For example, Public Health in the UK was responsible for live animal inspection, Food and Drug agents were tasked with ensuring food was free of chemical adulteration, while Ministry of Health had jurisdiction over imported and canned food, as well as food additives (Hardy 1999, 305). This lack of coordination meant that Public Health officials had a difficult time monitoring food related diseases and had to rely on public education as the best means to decrease food related illness and death. In fact it was only in 1938 that any move was made to acknowledge and track food poisoning despite the identification of salmonella enteritis in 1888 and several other bacilli including strains of E. coli and typhoid throughout the 1890s. This attitude to food safety – a form of liberal interventionism – placed the onus on the individual as consumers were given information and then expected to act in their own best interests (Draper and Green 2002).

In addition, the precautionary principle was not widely applied with respect to food safety in the interest of the public good as the UK Ministry of Health of the day erred on the side of industry and not the public. In a set of 1927 recommendations on food additives it was reported that, "Colour is frequently used to cover up objectionable or inferior materials, or to give a fictitious appearance, so that the objects so coloured masquerade for something which they are not" (Ministry of Health, 1924 in Cannon 1986, 1275). Despite the knowledge on the part of the Ministry of Health that colourants were at best disguising unhealthy food (in the early days) and more recently that the additives are known carcinogens, colouring additives were still used in the British food system into the 1980s. The lethargy in adopting more rigorous standards reflects both the disconnected facets of the various aspects of food regulations and the power of food processors – the latter point is explored more thoroughly in Chapter 7.

Following both WWI and WWII the UK focus shifted from a less interventionist attitude to the stimulation of food production and food security as the government aimed its sights on food quantity and quality. In the wake of both wars, the government focused on feeding its population. The production focus – although geared to the protection of the consumer as the recipient of adequate amounts of food – resulted in the increased industrialization of the food system as food production was structured to ramp up quantity. This exposed the UK food system to increased risk,

> During and after the war, there was a rapid expansion of mass catering, both in terms of feeding large numbers of people in canteens and restaurants, and in the sense of mass production of prepared foodstuffs...Egg-borne salmonella poisoning received widespread publicity, for example, when incidents were traced to the use of imported American powdered egg...Trade in both human and animal foodstuffs became internationalized, opening Britain to a large number of exotic salmonellas from all over the world. (Hardy 1999, 309)

The importation of contaminated egg powder from the US into the UK and other similar incidents provided a rationale for regulatory controls over food quality. In this way, global standards began to emerge as different trading countries sought to establish food safety guidelines.

Through the post-war period, the primary approach to food safety remained one of education as the government continued to practice "benevolent paternalism" as they provided guidelines for food safety and expected the public to take responsibility for their implementation (Draper and Green 2002, 614). The tension between paternalism and laissez-faire characterized UK food policy until the 1970s when heart health emerged as an issue of public concern. As a strand of what Lang and Heasman (2004) characterize as the nutritional nexus of social and bio-chemical mechanisms, heart health surfaced as one of the focal points of shifting policy. In the 1980s, with Thatcher at the helm, UK food took on a strong neo-liberal flavour. As part of this larger socio-economic transformation, health policy went from a social concern to an individual consumer responsibility, for as Thatcher espoused, society did not exist.

Under this new regulatory regime, food became more deeply entrenched as a 'commodity' and the human health associated with it, the responsibility of each and every consumer. In the case of heart disease, cancer and obesity (according to WHO heart disease was associated with 30% of annual deaths in 1999 while diet was linked to 30% of cancers, Lang and Heasman 2004), the public was advised about the benefits of low fat, high fibre, soy products, free radicals, Omega 3 and 6 and a wave of other food 'cures'. The attitude that the public should act unquestioningly upon government directives persisted well into the 1990s in the UK and is explored in greater depth in Chapter 5 when we discuss the Mad Cow crisis. But any expectation that individuals could fall lock-step into their new roles as masters of their own health was at least partly undermined by the 'SuperSize' portions and other marketing pressures that push eaters in the direction of consuming more convenience foods (Schlosser 2002). Not coincidentally, as well as being high in salt and sugar, these products are high value-added for food processors who have an interest in encouraging people to eat more.

By the 1990s as society witnessed the effects of the hollowing out of social supports and government as representing (if not championing) the public good, there was a gradual shift in the relationship between policy-makers and consumers. In the UK the pendulum swung back in the direction of a more balanced and inclusive policy environment with respect to food. This culminated partly in a change in consumers' roles beginning in the 21st century as more eaters became active citizens and informed eater (Gabriel and Lang 2006).

These changes were precipitated, to an extent, by food scares such as the BSE crisis and the introduction of GE food. Consequently, the UK government's approach to food safety has become increasingly transparent and participatory as a result of serious public scrutiny and pressure (Whatmore 2002). In the last few years, through the creation of progressive governance models, the UK has separated the public policy realm into more appropriate bodies for the protection, regulation and support for food, health, agriculture and commerce. At the same time, linkages have been established to permit coordination and collaboration between more branches of the government and the public and private sectors (Morgan et al. 2006). Next we will explore the development of the regulatory environment in the United States.

North American Food Regulation

In the United States a different story emerges.[2] As the links were being made between disease, bacteria and food at the turn of the 20th century, there was a genuine sense of urgency to impose scientific approaches and regulations onto the food preparation process in the United States. The watershed food regulations enacted in the US in 1906 were important and perceived to be necessary as they eliminated food borne illnesses as the leading cause of death at the time (Bobrow-Strain 2005, 5) and were seen as one way to generate a healthier, more productive America. This sentiment is aptly reflected in a 1905 *Good Housekeeping* article,

> National virility…depends upon individual health to such an extent, and this in turn is so largely governed by our food, that the healthfulness of foods is a matter of the most serious consequence to the nation. (Bobrow-Strain 2005, 12)

Shifts in meat processing and the rise of bakery-made white bread provide two excellent examples of the changes in the early part of the 1900s as they capture the main features of the shift from small to larger scale food processing.

At the beginning of the 20th century, Chicago was the centre of the meat processing industry where the one square mile meat district employed over 40,000 people. Animals were processed in four to five story slaughterhouses where live animals went in the top floor and sides of beef and other product came out the

---

2 This builds on the observations by Goodman (2003) and Allen et al. (2003) about differences between North American and EU food systems in a contemporary context.

ground floor. The meat was shipped across the US and to Europe. The workers were skilled but poorly treated, so that they commonly suffered lacerations, amputations and other injuries. In an attempt to expose the terrible working conditions, Upton Sinclair wrote an account of the industry entitled *The Jungle*. The stories in this well-researched documentary-style novel were so horrific they sparked a public investigation into working and sanitary conditions in the slaughterhouses,

> There were cattle which [sic] had been fed on "whisky malt" the refuse of breweries, and had become what men had called "steerly" – which means covered with boils. It was a nasty job killing these, for when you plunged your knife into them they would burst and splash foul-smelling stuff in your face, and when a man's sleeves were smeared with blood, and his hands steeped in it, how was he ever to wipe his face, or to clear his eyes so that he could see? It was stuff such as this that made the "embalmed beef" that had killed several times as many United States soldiers as the bullets of the Spaniards; only the army beef, besides was not fresh canned, it was old stuff that had been lying for years in the cellars. (Sinclair 1906, 2003, 102)

The public report produced in 1906 from the investigation prompted by Sinclair's book connected industrial food processing practices with food safety and raised public awareness about the health threats stemming from the industrial food system. The federal report led President Theodore Roosevelt to enact 'The Pure Food and Drug Act' to mitigate public health concerns. The federal investigation explained that workers laboured,

> under conditions that are entirely unnecessary and unpardonable, and which are a constant menace not only to their own health, but to the health of those who use the food products prepared by them. (Schlosser 2002, 152–153)

The history of bread in North America is an account of the industrialization, purification and sanitizing of food in American society (Bobrow-Strain 2005). The adage 'the greatest thing since sliced bread' speaks to the depth that the industrialization of this staple food product affected in North America. Aaron Bobrow-Strain's paper describes the ascendance of white bread, and the supposed progress from an impure, heterogeneous, home baked product in the early 1900s so that by the 1930s a new standard was achieved. According to Bobrow-Strain, during the first decades of the 20th century, adulterations and impurities in bread seemed to menace American society, threatening the physical and mental well-being of the population, impeding progress and undermining the greater good. By 1930, years of scientific attempts to manufacture pure and uniform bread had culminated in the legendary "invention of sliced bread", so that,

> it wasn't just the production of bread that had changed; bread itself had changed. By 1930, bread was white, sliced, and modern – and Americans loved it. (Bobrow-Strain 2005, 9)

The rise in popularity of sliced white bread was so widespread that by 1930, 80% of Americans purchased their bread from consolidated bakeries – a system of food provision that was non-existent 30 years earlier. What helped to spur on this transformation? Part of the answer lies in the perception about food and the view that food was sinful and needed to be controlled and made safer through scientific and regulatory intervention – enter Mr. Atwater.

At the turn of the 20th century, institutional influences in the guise of Wilbur Olin Atwater and the United States Department of Agriculture (USDA) led the charge against unhealthy food. Atwater, an employee of the nascent USDA, actively promoted the application of science to the preparation of food and championed the science of home economics. In 1895, Atwater introduced the first USDA food guide and spelled out the caloric value of different types of food. Atwater, of the opinion that health, body and morals are inextricably linked, espoused that people, "treat the body like a machine, and food solely as its fuel. Pleasure had most emphatically nothing to do with it" (Stacey 1994, 30). Atwater as a founder of nutrition studies in the US made several valuable contributions to and laid the groundwork for the field. His graduate training took him to Europe where he studied at experimental agricultural research stations where food production was being pioneered. Atwater imported the concept of agricultural research stations into the United States upon his return further linking agriculture with science. By 1894 Atwater had secured money to establish research stations and funds to conduct research on nutrition. Nutrition was key for Atwater as,

> . . .the intellectual and moral condition and progress of men and women is largely regulated by their plane of living; that the plane of their intellectual and moral life depends upon how they are housed and clothed and fed. (Kelley 1993)

Under Atwater's direction, four areas surfaced as important for nutritional research: the link between demographics and food consumption; the chemical make-up of food; the effect of cooking and processing on the nutritional value of food; and, optimal food requirements (Kelley 1993).

Meanwhile in the UK three similar proponents emerged – Rowntree, Wood and Hopkins. Taking up Atwater's 'minimalist' approach to nutrition that defined food requirements in terms of basic as opposed to optimal nutrition, a 1915 report assessing nutritional needs compared the human body to a steam engine with food as its fuel (Lang and Heasman 2004, 105). The conceptualization of food as fuel laid the groundwork for future tensions between different levels of government along with their increasingly encapsulated ministries, industry and public welfare interests. It also fed into the discourse of food as a commodity.

The reaction to the rivalry between the US dairy industry and the emergent oleomargarine manufacturing industry in the late 1800s is an early example of government intervening in agriculture. This instance is particularly relevant as it details the first case of a tax being used in the US as a competitive price barrier. In 1874, oleomargarine was developed in France by combining skim milk and processed animal fat. Food colouring was then added to make this whitish processed product more attractive to consumers. In the 1880s the US dairy industry successfully convinced several states to pass laws banning oleomargarine or to implement taxes to increase the price paid by consumers. According to Dupré (1999), 20 states required oleomargarine to be labelled while 7 banned selling the product altogether. Interestingly, the driving forces behind these initiatives were the cheaper cost to produce oleomargarine compared to butter and the oversupply of dairy products. Essentially then this battle was about the irrelevance of the cost of the raw material. As such, it helped to set the stage for subsequent attitudes about value-added versus commodity-value in the food industry.

Having secured protection at the state level, the dairy lobby pushed the issue into the federal arena. By 1886 they were successful and a law was passed that imposed heavy annual licensing fees of $600 on manufacturers, $480 on wholesalers and $48 on retailers, as well as a $0.02 per pound tax. This set a precedent for federal powers whereby the state imposed a tax as a protective measure for a specific industry, not as a revenue generating measure for the larger society (Gifford 1997). Inevitably, the tax led to fraud as oleomargarine manufacturers tried to sell their product as butter to avoid the taxes. This in turn led some jurisdictions to prohibit the use of colour to make oleomargarine less appealing or, as in the case of Vermont, New Hampshire and West Virginia, to add pink colouring to distinguish oleomargarine from 'natural' butter (Dupré 1999). The margarine taxes remained in place until WWII when dairy shortages caused the price of butter to increase dramatically, and the public demanded access to the cheaper oleomargarine. In 1949 and 1950 the House and the Senate repealed the federal law regulating the price of margarine. The ban on colour was lifted on a state-wide basis beginning in the 1940s with the final bans lifted in 1967 (Dupré 1999).

Canada followed a more extreme, industry-protectionist path. Based on a pledge to dairy farmers, oleomargarine was banned in Canada until 1949.[3] At that time, according to a CBC archived report,

> Household butter budgets could soon be cut in half now that the Supreme Court of Canada has lifted a ban on margarine. Manufacturers can legally produce butter substitutes and grocers can stock their shelves with margarine as of Dec. 14, 1948. Britain's colony Newfoundland now sells margarine for 35 cents a pound, nearly half of what a stick of butter goes for. But farmers aren't as thrilled as housewives. In today's CBC Radio broadcast, dairyman Earl Kitchen explains why. Kitchen is national secretary of the Dairy Farmers of Canada. He worries that if families begin using margarine instead of butter, farmers will raise fewer cows and there will be less milk. To make sure no one mistakes margarine for butter, margarine must be sold colourless in a plastic sack. Once at home, housewives press a tab on the sack to release the separately-packaged yellow dye into the white margarine. It takes 20 minutes of squishing to mix in the dye. (CBC 1948)

This brings us to a watershed period in food history – the period following WWII. As the world left the war behind, there was a reverence for the modernity, cleanliness, hygiene and convenience. At the same time 'deskilling' in the kitchen led to 'deterioration' in the quality of cuisine (Kingston 1994, 105). It was also the time when bland, processed food became the standard and people moved one more step away from natural, fresh food. The creation of food standards paved the way for the post-WWII era shift to an emphasis on modern, sanitary food as embodied in

---

3 There was a 4-year exception during this period due to reduced supply during WWI (Dupré 1999).

Andy Warhol's famous Fordist Campbell's Soup tin. This period in North American food production valued competence, scientific inquiry and the consolidation of mass production. Efficiency and hygiene were of prime importance so through the mid-1900s the emphasis moved to promoting convenience and 'snap, crackle, pop' (Stacey 1994: 15). Speed and efficiency were the bottom line as food was further transformed into a processed commodity, capitalizing on the view that food was fuel, and not a source of pleasure,

> Processed foods are often felt to be more healthful, and in this way more desirable, than foods consumed in a natural state, particularly since such foods are often presented with some sort of covering. Packaged, canned, wrapped foods, and bottled liquids at present have become endowed with the connotation of the *pure, the sanitary, and the healthful.* (emphasis added, Cussler 1952)

The 'scientific turn' that occurred over the course of the 20th century in part fuelled the changing relationship between people and their food. Government policy reinforced this shift as research programs became increasingly focused on and dominated by corporate research agendas. Scientific research is a huge lever in determining all conceivable outcomes from policy at the macro level through to the kind of food which is produced, and how it is processed, packaged and delivered, down to how food is or is not an actor at the micro level. Within this scientific regime research and intellectual property issues are key, and so will be discussed at some length.

Scientific research and policy in the United States, Canada and the United Kingdom

This brief history of scientific research in the US, UK and Canada describes policies that vacillate between research for the public good and research to benefit the corporate purse. From the early 1900s forward, a convergence emerges between the three countries in terms of policy as they all move to more applied research programs focused on biological and military technologies by the late 1980s. It is important to understand the evolution of public, private and government attitudes to research, technology and science as this informed the regulatory and policy environment that privileged private interests over public food safety concerns by the 1980s and beyond.

The United States

Science assumed a new role in the United States public and government psyches during WWII. Prior to WWII, university researchers and government labs operated independently. This changed in WWII as university scientists were called on to develop weapons, chemicals and products for the war effort. Emerging from the war, this created an extremely supportive climate for science generally and technology more specifically. The high level of support fostered the conversion of surplus production capacity for chemicals such as DDT and the need to find new applications for the chemicals that were used in wartime. As we discuss in more detail in Chapter 6, increasing population pressures, the need for more food and the opportunity for chemical applications for food production to control pests and weeds converged to establish the emerging chemical industry as an important pillar in the industrial food production regime (Levenstein 1993). Ultimately, this attitude led to the Green Revolution as the high tech, high input solution to world hunger.

As part of the new relationships being forged in R&D following WWII, the Department of Defense (DoD), the Atomic Energy Commission (AEC), the National Institute for Health (NIH) and the National Aeronautics and Space Administration (NASA) began to finance university research linked to the biological and health sciences (Wright 1994, 22). Between 1946 and 1956 federal R&D budgets grew from $917.8 million to $3.45 billion, and to $16 billion in the following decade (Wright 1994, 22). The Soviet threat, "proved more important for science, for it opened the government purse strings wider than ever before in peacetime and set off the mad scramble for weaponry that President Eisenhower would come to call the military-industrial complex" (Roland in Wright 1994, 22). Through the late 1940s and into the mid 1960s, the arms and space races received over half of government research dollars. On the health front, another important destination for public research dollars, powerful lobbying by the American Medical Association and proponents for cancer research steered health research dollars into biomedical research. This laid the foundation for the development of platform innovations and technologies needed for agricultural biotechnology. In 1966, arms and space accounted for 55.7 % of federal R&D and biomedical for 21%. As an OECD report stated in 1968,

> Almost all [U.S.] science and technology activities are part of planetary competition, and are expected to reinforce not only the country's military position but also its economic potential, social equilibrium, and international prestige. (OECD in Wright 1994, 27)

However the enormous amounts of money allocated to scientific R&D were not handed over without some critical questions from various US citizen groups. The huge US research budgets (in 1963 the US spent $12.5 billion on R&D) came under scrutiny as public priorities shifted in the late 1960s (Wright 1994, 24).

All was not straightforward for the scientific community as the world transitioned through the 1960s and overall public skepticism emerged. Rachel Carson's the *Silent Spring* opened the doors for scrutinizing the global reach and persistence of chemicals. And, as humans saw the earth from space for the first time, an awareness about the fragility of 'spaceship earth' emerged (Fuller 1963). The escalating war in Vietnam provoked strong public reaction and served to focus radical public attention on the weapons – particularly the chemical weapons – that resulted from science and technology. The Kent shootings of May 4, 1970 when four students protesting the Vietnam War were killed and nine others injured are symbolic of the division between state and citizenry at the time. A complex relationship emerged among the scientific community, government, the military and the public as the US questioned its role in Vietnam. This self-examination provoked questions about the role of scientific innovations such as the chemical Agent Orange and prompted the administration to reexamine (and repackage) its scientific R&D. Nixon, for example – in an attempt to deflect public opinion away from events in Vietnam – declared war on cancer in his 1971 State of the Union address,

I will also ask for an appropriation of an extra $100 million to launch an intensive campaign to find a cure for cancer, and I will ask later for whatever additional funds can effectively be used. The time has come in America when the same kind of concentrated effort that split the atom and took man to the moon should be turned toward conquering this dread disease. Let us make a total national commitment to achieve this goal. (National Cancer Institute 2007)

A first step in realizing this goal was the conversion of the Army's Fort Detrick, Maryland biological warfare facility to a cancer research center. By the 1980s, "The formulation of policy for the development and control of science and technology remained the prerogative of government and private industry" (Wright 1994, 43).

The United Kingdom

In the UK the separation of university funding from direct government influence was established in 1919 and continued through WWII until the 1960s. The division successfully kept publicly funded scientific inquiry removed from private interests. This goal was met through independent Research Councils that allocated funds to universities for research. In the 1960s the separation between independent, publicly funded research blurred and the relationship between science and society shifted toward an emphasis on accountability similar to the post WWII US system. The economic challenges that plagued the UK in the aftermath of WWII continued into the 1960s and meant that R&D programs could not grow as quickly as those in the US.

   At the same time, a partisan pattern emerged. On the one hand Conservative governments cut funding. On the other hand, Labour governments moved money from military spending to health-based research. Through the 1960s, the UK struggled to develop applied research programs to buoy the sagging economy, offer alternatives to old industries and keep their talent from fleeing to the US (Wright 1994, 31–34). The Conservative government led to victory by Edward Heath in 1970 determined that Government should establish research directions and scientists should fulfill these needs. This signaled,

> an end to an era of science policy committed to the principle of scientific autonomy and the onset of a new era in which the goals of research and development were directly defined by the state and progress toward their realization was used as a criterion for support. (Wright 1994, 36)

As the 1971 Rothschild report stated, "The country's needs are not so trivial as to be left to the mercies of a form of scientific roulette" (Rothschild 1971 in Wright 1994, 36). On the science policy-R&D side, the emphasis on accountability opened the door for the increasing support of advanced biomedical innovations such as biotechnology as these programs could be measured and quantified in terms of their supposed relevance and applicability. In 1976, the government formed the Advisory Committee for Applied Research and Development (ACARD). With strong industry representation, the committee recommended, a greater emphasis in Britain on the sciences related to manufacture with special attention to the relationship between production processes and the design, quality and reliability of products, as well as to marketing. (Wright 1994, 60–61)

Thatcher's government reinforced this direction. And so, the UK moved in a firmly interventionist direction by the 1980s with respect to publicly funded R&D as it set its sights on applied research tied directly to industry needs.

Canada

Canada presents a hybrid story of the evolution of scientific research and public policy, ending up somewhere between the US and UK accounts. In the following sections we also explore the links between IP and government research programs.

The 1800s, focused on developing the natural resource capacity of the country. At the urging of the British government, the Canadian National Research Council (NRC) was founded in 1916 as a way to, "offset German dominance in science and technology" (Atkinson-Grosjean et al. 2001, 7) and to develop industrial research capacity (Fisher et al. 2001). At the outset, the NRC determined the research capacity of private interests to be so minimal, it established its own research facility in Ottawa. The creation of the NRC cemented nationally funded research and private interests together from the outset of Canadian federal R&D policy. At the same time, a small privileged network of scientists sat atop the nexus between the academy and industry. As a result an elite group in the Canadian research establishment made important decisions about the direction of Canadian R&D policy well into the 1960s. For example in an interview about the creation of a nuclear research unit during WWII, scientist C.J. Mackenzie explained the ease with which the nuclear research program got its start in Canada,

> It was surprisingly easy. In those days the NRC reported to C.D. Howe [then Minister of Department of Trade and Commerce].... C.D. was a particular friend of mine.... We all went to C.D.'s office and discussed the idea with him. I remember he sat there and listened to the whole thing, then he turned to me and said: 'What do you think?' I told him I thought it was a sound idea, then he nodded a couple of times and said: 'Okay, let's go'. (Lee 1961 in Atkinson et al. 2001, 2)

Public and private interests were shifting through the 1960s. Atkinson-Grosjean et al. explain,

> The funding of basic (public) research was justified in terms of its eventual but distant (private) payoff which [sic] would, in turn, generate future (public) returns in the form of tax revenues, employment, and technological innovations. Because of the power of resource allocations from the state, the linear model became, in effect, a self-fulfilling prophecy, constructing real divisions from imagined boundaries. (Atkinson-Grosjean et al. 2001, 4–5)

During the 1960s visions of and for the NRC oscillated between a focus on unfettered independent basic research on the one hand, and research targeted to and directed largely by the needs of private industry on the other. By the 1970s, the beginnings of a foundational shift were underway as the direction moved firmly in favour of corporate interests. The Lamontagne Committee report recommended,

> The organizational structure for the R&D system should be based on economic forecasting and planning for future industrial needs, giving industry priority while taking into account the needs of science and Canadian society. (Atkinson-Grosjean 2001, 13)

Gradually, public research dollars were leveraged for private purposes. The shift in the ownership of the intellectual property (IP) for soybeans in Canada is a case in point and is explored later in the chapter. The transition is especially relevant in the contemporary context of food fears as corporations replaced governments as the drivers of research, and increasingly lobby government about what they define as appropriate regulatory environments. As a result, regulatory regimes are constructed with corporate needs in mind.

By 1987, the pendulum had swung firmly in the direction of national industry priorities and international competitiveness. The federal mega-ministry Industry, Science and Technology Canada (ISTC) was created to hand out money and oversee a national research network. Funding for research by non-industry players such as universities was then tied directly to matched funding from private interests. This move was in step with initiatives in other OECD countries as the developed world ramped itself up to compete in the New Economy. The introductory description for the OECD[4] 1997–1999 National Innovation Survey explains,

> Innovation is seen to play a central role within the knowledge-based economy and this has led to greater policy attention being paid to the processes of technological innovation and diffusion by firms. Innovation surveys have been developed in response to the need for reliable and systematic data for the design, monitoring and evaluation of policies aimed at promoting technological innovation. (Muzart 1999, 5)

Since the development of the 1987 Science and Technology policy, Canada has adopted a strong focus on funding economically relevant research (Godin et al. 2002).

In step with US policy over the last fifteen years, Canadian government funding for agricultural research has been largely tied to matching dollars from industry although lagging behind US spending levels. As substantiated in the next section, all of this – the requirement for matched dollars, the reduced amount of available research dollars and the connection to private research – creates a policy environment that favours private over public interests. This trend contradicts the stated goals of the federal government so that:

> New knowledge needs to fuel Canadian innovation that, in turn, affects every aspect of food and non-food production, changing the way Canadians grow, process, preserve, transport, distribute, and use the products derived from agriculture. In other words,

---

4 OECD member countries in 2007: Australia, Austria, Belgium, Canada, Czech Republic, Denmark, Finland, France, Germany, Greece, Hungary, Iceland, Ireland, Italy, Japan, Korea, Luxembourg, Mexico, Netherlands, New Zealand, Norway, Poland, Portugal, Slovak Republic, Spain, Sweden, Switzerland, Turkey, United Kingdom, United States.

new discoveries and their application are crucial to ensuring Canadian farmers and the Canadian public benefit from Canada's natural advantage, i.e. its ability to produce food and an ever-increasing range of non-food products from the land. Examples of these new applications include new bio-materials, bio-medical and bio-health products, bio-fuels, bio-energy, bio-chemicals, and bio-pharmaceuticals. (AAFC 2006, 1)

The emphasis on food for fuel and pharmacy exemplifies the tensions that emerge as private displaces public goals. Given the desire to stimulate economic benefits through the creation of new knowledge, it is important to understand existing and future innovation capacity to achieve this goal. As an example of the change in research directions in North America and the UK and the ascendancy of private IP rights over public ones, the next section presents results from a four-year research project that explored dimensions of innovation for the agro-biotechnology industry in Ontario.

Agro-biotechnology case study Contradictions abound within the IFS as different interests compete for market share, research funding and policy support. This section describes the importance of basic research for agro-biotechnology and provides comments about the role of intellectual property rights in the context of basic research. These observations offer insights about access to corporate controlled IP and the extent to which this affects public innovative capacity. The research allows us to understand some of the more nuanced implications of public policy and the way this affects public control over food research geared toward the public good. It also informs how government spends food research dollars, and how it makes decisions about food safety. This in turn influences the level of scrutiny by government over the food system and the quality of food that people eat. It helps to explain how, in the face of the overwhelming majority of Canadians wanting food with GE content labelled appropriately, there are still no labelling requirements. The findings come from a case study of agro-biotechnology innovators in south western Ontario in the springs of 1998 and 2001 (Blay-Palmer 2003). Sixty interviews were conducted with innovators in government, private and university research labs. Policy-makers were also interviewed to provide a context for innovation.

It is well understood that basic, curiosity driven research is critical for innovation as it leads research in new and unexpected directions (Nelson 2004; Doern 1999; Pavitt 1991). This is the case in Canada as highlighted by two examples reported during in-depth interviews. In the first case a public research lab discovered a new biotechnology identification method. This innovation replaced an existing method that was owned by private interests and was becoming expensive to access. Although the development of the identification method was not the goal of the public lab's initial research, the project produced the unexpected innovation that offers all researchers an alternative method that is now readily available in the public domain. Having this IP in the public domain means it will be more readily available and at a more reasonable cost than the previous IP developed by private interests. This will help to fuel more research and is a clear example of the unexpected and important spin-offs from publicly funded, basic research (Doern 1999; Rey and Winter 1998).

The second example of the benefits of basic, exploratory innovation is the creation of a cold-tolerant variety of soybeans through conventional breeding practices in the 1970s. While engaged in traditional breeding of another trait but using germ plasm from cold tolerant eastern European plant varieties, university and government breeders in Ontario noted that some of their plants were able to withstand cold conditions better than others. The subsequent development of a cold tolerant variety of soybeans spread the growing range for soybeans into eastern Ontario. The effect of this innovation for farmers was profound as it provided them with a crop to alternate with corn. Valuable new markets for high quality soybeans for tofu made this a particularly interesting option. For consumers, it provides a local source for quality, oilseed based protein. The two examples of unexpected, serendipitous research results provide evidence of the merits of unfocused, public basic research for food production. In both cases, public interests were well served by an investment in public research.

The next part of the soybean research underscores the importance of basic innovations for future innovative capacity and the spin-off benefits for private companies. It also tells the story of the domination of the ownership of agricultural IP and the rights to seed stock by MNCs and raises questions about public versus private control over various aspects of the food system. Returning to the cold tolerant soybeans, the uptake by farmers was rapid as they embraced the new opportunity. This was so much the case that the increase in soybean growing area attracted the attention of MNCs and they began to invest in soybean breeding. The improved commercial viability of soybeans translated into a shift in the ownership of soybean varieties between 1975 and 1998. In 1975 90% of soybean varieties were publicly owned while private interests controlled only 10% of the soybean IP. In a complete reversal, by 1998 only 10% of the varieties were publicly owned (Smithers and Blay-Palmer 2001). The ramifications of this shift are profound as corporate concentration means more seed resources are owned by private corporations.

The development of soybean varieties over this period coincided with GMO research and, importantly, the release of RoundUp Ready Soybeans. In a controversial move to protect their IP, MNCs required farmers who purchased RoundupReady soybean seed to sign Technical Use Agreements (TUAs). Through TUAs farmers agree not to save their seed to replant the next year and obligates them to buy seed every year. As a result, seed companies were able to justify increased investment in R&D for soybeans whose seed could otherwise be harvested and replanted from year to year. Commenting on the creation of the GMO RoundUp Ready soybeans, experts indicated during interviews that as they could then recoup their R&D investment through IP and the payment of TUA fees, more innovation occurred for this crop. According to one key informant, "there is increased reinvestment in R&D where protection exists". The case of soybeans also underscores the value of this research as a building block in the cumulative innovation process (Nelson 2004). Expert informants explained that the protection of IP is imperative for the *continuation* of capital-intensive research and development.

However, as biotechnology research is largely incremental and cumulative (Horbulyk 1993), the first discovery can begin a "chain of dependency...in a whole cascade of linked products" (Cassier 2002, 90). In this way the patenting of core processes and platform technologies can increase the cost of developing next generation innovations for other researchers who need to buy access to the new technology (Marshall 2000) – for example, the chain that was unintentionally interrupted by the creation of the new identification technique in the Canadian research lab. As scientists look down the road to the production stream, if there are too many 'development tollbooths' in the form of licensing and royalty fees (Heller and Eisenberg 1998), innovators will not pursue a particular research direction (Merz 2000). This prevents research from unfolding along a serendipitous path, impedes overall innovation and slows basic research (Nelson 2004; David 2001). A lack of access to corporate controlled IP can also impede innovation as some companies refuse to license technologies, or limit access as in the case of the Dupont oncomouse. These 'tollbooths' can be so onerous they derail downstream innovation. Dupont's use of IP – for example, the 'oncomouse' and Cre-loxP (Marshall 2000) – is a case in point. Dupont, in an effort to capitalize on their IP, built into their initial use agreements for the Oncomouse and Cre-loxP – veto rights over the development of new technologies that emerged from work involving their IP. Dupont's goal was "to leverage its proprietary position in upstream research tools into a broad veto right or revenue generation for downstream research and product development" (Heller and Eisenberg 1998, 699; see also Marshall 2000). Dupont then used its control over platform technologies to block innovation of other technologies that could build on their innovations.

Clearly then, access to IP is an important consideration for the direction of future innovation as the rights of the patent holder/ licensee can take precedence over the desire to disseminate information and the interest of the social good (Foray 1995; Gotsch and Reider 1995; Scotchmer 1991). David (2001) provides an example of the conflict that results from limiting public access to IP. In his examination of the digital data industry and the European Community's Database Directive, David explains that access to knowledge that belongs in the public domain is under threat,

> an unchecked bias towards expanding the domain of information goods within which private property institutions and market mechanisms flourish, is steadily encroaching upon the domain of public information. In doing so, it has tended to weaken, and may in the end, seriously undermine those non-market institutions which historically have proved themselves to be especially effective in sustaining rapid growth in the scientific and technological knowledge base that is available to be exploited. (David 2001, 3)

David cautions that an unintended boomerang effect may – and we have, in fact seen this in operation in the forgoing examples – return to harm those institutions that formulated the laws and policies in the first place as IP protection could deny them access to IP and the seeds needed for food (Merz 2000; Cassier 2002). One public innovator explained,

> It is not always possible to gain access to the best IP or reagents to do the job. Companies often put unreasonable restrictions on access. Public institutions and government laboratories need sufficient resources to acquire or develop enabling technologies that allow them to compete with private companies.

There are also access issues between private firms (Scotchmer 1998). Another key informant explained that between companies "with competitive intelligence, people play their cards a lot closer to their chest". IP raised contradictory issues for innovators working within MNCs. In some cases, it was perceived to foster innovation. For these innovators it "gives you currency, [you] can trade it for what you need. Today cash is probably the least interesting piece of currency you can offer when interacting with other companies". In other cases, it produces an atmosphere where people are protective of their innovations, "driving IP into dark corners as IP has value". Even when IP is accessible, it may require, "multiple agreements to conduct research". Negotiating these agreements and putting them in place can delay collaboration and access to funding, slowing the research process and increasing transaction costs. Ultimately this threatens what does and does not get developed as food.

The sources of funding can have an effect on the outcome of projects. In Canada, funding for university research comes from both provincial and national sources. Government research dollars are awarded to projects that solicit matching dollars from the private sector. Inevitably, this spins research projects in the direction of applied research (Bordt and Read 1999; NBAC 1999). Consistent with research by Toole (2000) and Scott et al. (2001), one person interviewed explained that to have business driving research directions is bad for basic science, "There is always a conflict between good science and the need to address industry issues". Many at Agriculture and Agri-Food Canada (AAFC) indicated that there is an on-going compromise between conducting the research they need and the research they can get matched funding for. As one innovator explained they must, "fit ideas onto the list [of acceptable projects]". The pull in the applied direction is clear so that, "Funding opportunities require industry support however we need to have basic research, not just applied with companies. It [research] should not just be driven by the market".

Agro-biotechnology and the public good in Canada   All of this plays back into the larger discussion about food systems, food safety, and the attitude of the public toward food. Recalling the general question concerning public versus corporate priorities and the conflicted atmosphere this creates as private interests supercede public ones, it is important to examine the public's position on government-supported research – in this case, public opinion about agro-biotechnology that, in the form of Genitically Modified Organisms (GMO) products, pushes relentlessly onto the dinner table. In polls about food safety, consistently 85% or more Canadians want food containing GMOs to be labeled (e.g. DECIMA 2004, Environics 2001). In a 2004 document prepared by Agriculture Canada, it was reported that 77% of Canadians have concerns about the safety of food that contains GMOs – that is food created through the application of agro-biotechnology. Despite such clear messages from the public and experts about their concerns (Royal Society of Canada 2002), the federal government continues to provide tens of millions of public research dollars every year to food related biotechnology research. In the case of agro-biotechnology in Canada, there is a disconnect between the public good and government policy.

Meanwhile, more effective consumer opposition in Europe, New Zealand and Japan to GMO food has an impact on public researchers in Canada. Some interviewees reported that research projects had been scaled down or cancelled as a result of this uncertainty[5] and resistance from key export markets. Most innovators indicated that consumer questions and concerns are reasonable and should be addressed. One researcher expressed concern about popular opinion indicating that there is, "some uncertainty about GMOs at all levels. First products were brought to market too quickly. [There] could always be increased research. We need to do open research…information used in government assessment should be available to public researchers". Another scientist explained, "In general people are respectful of science, although there are some questions about GMO and the extent to which it serves industry instead of people. These are reasonable questions".

This case study reveals several areas of concern with respect to food systems including the decline in innovative capacity and conflicts over IP. But, most importantly it brings to light the glaring conflict between the public good and the private purse. Despite the need for public scrutiny of the food system, the research agendas and policies of Canada established a research environment that privileges corporate over public interests and is an example of the tone in the US and in some respects the UK research regulatory environment at the time. Given that food is the 'commodity' under discussion, a loss of control by public interests can have far reaching effects as the food we eat is increasingly controlled by MNCs.

Converging R&D policy

By the late 1980s the UK, US and Canada had adopted similar policies for scientific R&D. On the trade front, under Thatcher, Reagan and Mulroney respectively, markets were liberalized as free trade deals were negotiated. While R&D policies were changing into the 1980s, industry and the scientific community were also experiencing a transformation with respect to their approach to R&D. Emerging from the 1960s and 1970s, the scientific community and industry in developed countries determined that: 1) they needed to have more input into the policy-making process in order to ensure that regulations were easier to work with; and, 2) science and development was packaged in a more positive light for public consumption as one way to ensure continued government funding (Wright 1994, 52). According to Wright's (1994) assessment of the 1970s business climate, industry felt over-regulated and was drowning as a result. She cites one executive as saying, "My industry is regulated up to its neck. You are regulated up to your knees. And the tide is coming in". The executive then goes on to predict that industry would be, "engulfed by the rising sense of entitlement" and the public's say in research and development priorities (Wright 1994, 52).

The impact of the 1980 Bayh-Dole Act in the US that allowed universities to own patents created from federally funded research helped to cement R&D relationships between universities and private companies. This built on the trend begun in the 1970s for companies to contract research to university labs. The first such contract between Harvard Medical School and Monsanto gave the university $23 million

5 This is consistent with research in other countries, for example the May 2006 closure of the Cambridge research branch of Biogemma and the loss of 37 jobs (Agrafood Biotech 2006).

over ten years for anti-tumor research, while Monsanto retained first patent rights on related discoveries (Culliton 1977). Strong links and savvy business management created a new environment for R&D.

At the same time, responding to potential resistance from a public whose faith had been shaken by – for example – the Chernobyl disaster, strategic decisions were being taken in the fledgling biotech industry. As Fleising (2002) remarks in a paper published in *Trends in Biotechnology* entitled 'The legacy of nuclear risk and the founder effect in biotechnology organizations',

> The founders and builders of biotechnology, both on the business side and on the regulatory side, and also including *ad hoc* citizens' groups against recombinant technology were socialized into an ethic of participatory technology during the height of the environmental movement. These founders either participated directly or were in a position to be informed and/or involved in the collective will that convinced Congress in 1968 to pass the National Environmental Policy Act (NEPA). (Fleising 2002, 159)

So while the public made gains in getting NEPA enacted, industry was able to block the application of NEPA to "halt gene-splicing technology" (Fleising 2002, 159) so that,

> In the wake of the Chernobyl nuclear accident and a decline in the public trust of science, the founders of modern biotechnology recognized the strategic importance of risk assessment and regulatory affairs. In an effort to avoid the demonization that was attached to the nuclear industry, the pioneers of modern biotechnology delegated authority for regulatory negotiation and risk management to senior positions in the firm. At the same time, the Biotechnology Industry Organization was handed great latitude and trust with making public pronouncements on issues of bioethics and public policy…An organizational culture sensitive to the power of public perceptions of technology and individuals socialized by environmental activism to a culture of expertise for negotiating risk, gave biotechnology the regulatory expertise required to establish the economic space that it now occupies. (Fleising 2002, 156)

Learning from the 'demonization' of the nuclear industry, the biotechnology industry deliberately established a senior level cadre of experts able to manage public perception. This move in the early days of the industry established a PR beachhead and facilitated the creation of the current 'economic space' that the biotech industry now 'occupies'.

Although beyond the scope of this book, there has also been a trend in the last decade to multi-lateral trade agreements as a complement to firm-led innovation and National Innovation Systems. This level of intra-national trade agreements adds another level of complexity to the current industrial food regime. As part of these initiatives, research policy and IP agreements through the 1980s and 1990s were streamlined so that global standards emerged. The negotiation of multi and bilateral agreements such as the international Treaty on Related Aspects of Intellectual Property Rights (TRIPS) and the UN World Intellectual Property Office (WIPO) contribute to a more porous global economy for multi-national corporations (MNCs). The transformation of the General Agreement on Tariffs and Trade (GATT) through the Uruguay round of talks into the World Trade Organization (WTO) was a landmark development in 1995. The increasing complexity of the innovation and regulatory environment further legitimates the role of science in the 'creation' of food innovations and adds substantially more distance between people and their food.

Diverging Approaches, Contemporary Food Regulation, Trade and Food Fears

Despite all the forgoing discussion of the globalized, industrial food system, food scares do close borders. The UK and Canada discovered this immediately when they reported their first cases of BSE/Mad Cow. The increasingly globalized nature of the food industry has created new markets and new financial opportunities, but has simultaneously resulted in the harmonization of food regulations and standards. According to the Economic Research branch of the USDA,

> In the past, outbreaks were mostly acute and highly local and resulted from a high level of contamination. Now, we see relatively more outbreaks from low-level contamination of widely distributed commercial food products affecting many counties, States, and nations. This development has been attributed to changes in food production and distribution and to the growth of international trade borders. (2002, 1–2)

In some cases this has resulted in more stringent regulations, in other cases this has created disputes. Two case studies will be explored to illustrate some of the larger conflicts that inform this debate. It is interesting to note first that the context for these conflicts is largely the result of the size and scope of the global trade in food including issues of jurisdiction, regulation and food safety. Second, is the notable lack of precautionary principle thinking that was officially invoked in the EU in 2000 with the adoption of a Communication on the precautionary principle by the European Commission. Free-trade advocates see the precautionary principle as a trade barrier and advocate for 'objective' scientific standards for control and risk assessment (e.g. Post 2006). Others applaud the EU for adopting a moral guide as a, "complement to science, to be invoked when a lack of scientific evidence means that outcomes are uncertain" (Carr 2001, 31). First, we will examine the EU ban on animal growth hormones as a test case for dispute resolution, the application of the precautionary principle and related sovereignty with respect to perceived food threats. Second, we will assess the Hazard Analysis and Critical Control Point System (HACCP) as the basis for local enforcement of globally accepted standards for food management processes. Both will serve to provide examples of weaknesses in the international regulation of food.

The Growth Hormone Controversy

Growth hormones are currently used in up to 80% of beef produced in the US (Center for Food Safety 2007). The hormones used can be broadly described as growth hormones that result in more milk (in the case of rBGH) or more tissue (i.e. meat) in the case of oestradiol 17-ß, testosterone, progesterone, trenbolone acetate, zeranol and melengestrol acetate. In 1988, the EU banned the importation of meat containing these additives. In 1997 the WTO ruled in favour of the US and Canada on their challenge to the EU right to ban beef imports on the basis of growth hormone content. The ruling was made in consideration of Agreement on the Application of Sanitary and Phytosanitary Measures (SPS).[6] The appellate branch of the WTO overturned the rulings but left open the opportunity for the EU to provide evidence of human health threat. Accordingly, in 1998 the EU initiated 17 research projects through its Scientific Committee on Veterinary Measures relating to Public Health to determine the health implications for growth hormones fed to beef animals including, "toxicological and carcinogenicity aspects, residue analysis, potential abuse and control problems, [and] environmental aspects". The findings, confirmed again in 2002, were that,

> As concerns excess intake of hormone residues and their metabolites, and in view of the intrinsic properties of hormones and epidemiological findings, a risk to the consumer has been identified with different levels of conclusive evidence for the 6 hormones in question.
>
> In the case of 17-β oestradiol there is a substantial body of recent evidence suggesting that it has to be considered as a complete carcinogen, as it exerts both tumour initiating and tumour promoting effects. The data available does not allow a quantitative estimate of the risk...
>
> For all six hormones endocrine, developmental, immunological, neurobiological, immunotoxic, genotoxic and carcinogenic effects could be envisaged. Of the various susceptible risk groups, prepubertal children is the group of greatest concern. Again the available data do not enable a quantitative estimate of the risk.
>
> In view of the intrinsic properties of the hormones and in consideration of epidemiological findings, no threshold levels can be defined for any of the 6 substances. (European Commission Health & Consumer Protection 2002, 6)

Despite these findings the US continues to use growth hormones in beef while the EU continues to ban its import. As a result the EU is subject to trade penalties amounting to hundreds of millions of US dollars ($117 million in 1997 alone) (Iowa Farm Bureau, n.d.). Clearly the evolution of different regulatory bodies has resulted in different approaches to food safety and the protection of public health – and this creates different cultures with respect to food fears.

Third Party Food Safety Certification: Hazard Analysis and Critical Control Point System (HACCP) et al.

The long distances that food is shipped and the many handling points along the way combined with numerous food scares precipitated a range of third party certification systems as national and international governments and organizations attempted to

---

6 Established under the World Trade Organization in 1995 recognizing the need for transparency, equivalence, science-based measures and harmonization of standards at a global scale (Buzby and Unnivher 2003).

control riskier behaviours throughout the food chain; an IFS 'system' response. One of the most widely used is Hazard Analysis and Critical Control Point System (HACCP). Initially the brainchild of the National Aeronautics and Space Agency (NASA), the US Army Natick Laboratories[7] and the Pillsbury Company, HACCP was created to improve food quality for the space program. Gradually, the HACCP program has been adopted as the standard for food handling throughout the developed world. In January 2006, the European Parliament on the Hygiene of Foodstuffs through Regulation 852/2004 required all food business operators to implement HACCP. Various stages and aspects of HACCP have been integrated into the Canadian food industry standards and regulations since 1998. They are also central to the Codex Alimentarius, part of the FAO/WHO Food Standards Committee whose role is to implement:

> ...the Joint FAO/WHO Food Standards Programme, the purpose of which is to protect the health of consumers and to ensure fair practices in the food trade. The Codex Alimentarius (Latin, meaning Food Law or Code) is a collection of internationally adopted food standards presented in a uniform manner. It also includes provisions of an advisory nature in the form of codes of practice, guidelines and other recommended measures to assist in achieving the purposes of the Codex Alimentarius. The Commission has expressed the view that codes of practice might provide useful checklists of requirements for national food control or enforcement authorities. The publication of the Codex Alimentarius is intended to guide and promote the elaboration and establishment of definitions and requirements for foods, to assist in their harmonization and, in doing so, to facilitate international trade. (Food and Agriculture Organization/ World Health Organization 2005, Preface)

HACCP is now an internationally applied tool that standardizes food safety procedures. It provides a system of verification at multiple points in the food system from field to fork. HACCP identifies critical points where food is perceived to be susceptible to contamination and establishes best practices for ensuring food safety. The goal is to provide a level of confidence as, according to the Codex Alimentarius,

> People have the right to expect the food they eat to be safe and suitable for consumption. Foodborne illness and foodborne injury are at best unpleasant; at worst, they can be fatal. But there are also other consequences. Outbreaks of foodborne illness can damage trade

---

7 Natick is still central to US military provisioning. It is described on the US Military web site as:

"U.S. Army Soldier Systems Center – Natick, MA
More Than 50 Years of Protecting Those Who Serve
We're the science behind the Soldier
Food, Clothing, Protective Equipment, Parachutes, Shelters and Medical
Everything the Soldier wears, carries and consumes is either designed, developed or integrated here
From basic research through customer service and repair parts support - our focus is on supporting the Soldier ...
Statistics
Employees: 2,000 civilians, military and contractors
Combined annual budget: >$1B"
(http://www.natick.army.mil/, accessed online 30 April 2007)

and tourism, and lead to loss of earnings, unemployment and litigation. Food spoilage is wasteful, costly and can adversely affect trade and consumer confidence.

International food trade, and foreign travel, are increasing, bringing important social and economic benefits. But this also makes the spread of illness around the world easier. Eating habits too, have undergone major change in many countries over the last two decades and new food production, preparation and distribution techniques have developed to reflect this. Effective hygiene control, therefore, is vital to avoid the adverse human health and economic consequences of food-borne illness, foodborne injury, and food spoilage. Everyone, including farmers and growers, manufacturers and processors, food handlers and consumers, has a responsibility to assure that food is safe and suitable for consumption.

These General Principles lay a firm foundation for ensuring food hygiene and should be used in conjunction with each specific code of hygienic practice, where appropriate, and the guidelines on microbiological criteria (Food and Agriculture 1999, 3).

The relative importance of individual health versus the goal of ensuring proper trade flows for industry are presented as important goals for the FAO/WHO – indeed, one could argue that more consideration is accorded to issues of trade than to individual human health in this document. As these regulations are increasingly the international standard, the paperwork and scrutiny to comply with these standards privilege large, globally focused companies over small processors.

Conclusion

Chapter 2 described the evolving regulation of food systems in the United Kingdom, the United States and Canada since the 1800s. We have also surveyed the science and research policies of these countries until the late 1980s. By that time, there was convergence on matters of:

1. Public support for privately directed research in the food and biotechnology industry.
2. Strong institutional relationships between government and the food industry with diminished concern for public interests.
3. A scaling up of regulations to support industrial scale food processors and a North America shying away from precautionary principles supported in the EU.

The regulation of food supported a highly industrialized global approach to food provisioning. The institutional norms favoured technology and mass production.

In the next chapter we consider the marketing of processed food as another dimension in the emergence of food fear as a defining feature of the late 20th century food industry. With the food processing industry firmly entrenched in the North American landscape by the early 1900s, the marketing of food became increasingly important. The industry turned to marketing as the way to create and then fill market opportunities (Nestle 2003). According to food historian Harvey Levenstein (1994), when you have a culture in which food is the object of fear and loathing as well as love, there are people who are going to discover innumerable creative and inventive ways of exploiting these fears.

The next chapter makes some of the links between groundbreaking changes in food retail and marketing and the way this shapes the current food system.

Chapter 3

'It's All About the Sizzle'

Chapter 3 examines the connections between food processing and the rise of product branding and marketing and the way food processors and retailers use these tools to reassure consumers about food reliability and safety. One goal of this chapter is to describe the modes of regulation in retail to better understand how marketing is used to construct consumer's attitudes toward particular food products. In some cases marketing and branding may be used specifically to assure consumers that their food is safe; in other cases, marketing and branding are used to convince consumers that individual products are healthy, free of fats and contaminants, and even good for the broader natural environment. The case studies of brand names such as Kellogg's Corn Flakes, St. Michael's and President's Choice illustrate innovative marketing strategies that have been used to build consumer confidence in processed food.

This chapter covers the period from the early 1900s to the beginnings of the more explicit and mass food scares of the 1980s and 1990s. The end of the timeline for this chapter marks a convergence of three contemporary food-related issues: (1) a renewed consumer questioning regarding the ability of the conventional, agro-industrial food system to provide reliable, healthy food; (2) the constructed trust in branded products through extensive and in some cases aggressive marketing to adults and children; and (3) the growing consumer need to be reassured about food safety in the wake of the increasingly complex food system and the proliferation of food scares. As we shall see by the end of the chapter, much of current food marketing exists to create market share for highly processed food products. Marion Nestle describes the fierce competition and marketing in the food industry, and the way the oversupply of food production in the US keeps commodity prices depressed for farmers,

> ...[o]verabundance alone is sufficient to explain why the annual growth rate of the American food industry is only a percentage point or two, and why it has poked along at a low level for many years. It also explains why food companies compete so strenuously for consumer food dollars, why they work so hard to create a sales friendly regulatory and political climate, and why they are so defensive about the slightest suggestion that their products might raise health or safety risks. (Nestle 2004, 13)

At the end of the day, the most lucrative way for processors to increase profit is to add value through product differentiation and market segmentation. Otherwise, they are engaged in a race to the bottom as volume becomes the driving force behind increasing profits. As we shall see later in the chapter, Loblaws and Marks and Spencer (M&S) have realized the benefits of offering differentiated product to their customers.

Kellogg's Toasted Corn Flakes

The roots of food commodification can be traced back to the trading of spices, teas and other exotic food in biblical and then colonial times. However the process began in earnest with large-scale, industrialized food processing in the late–1800s. In an English Food Heritage book describing food and cooking in 19th century Britain, the kitchen of the 1890s is compared to a kitchen of the 1980s,

> The cook's store cupboard and larder in the middle class city home of the 1890s was as full of packets and cans as any housewife's now. With the help of bottled sauces, canned vegetables and fruit, and essences, she could choose between as many flavours (if less subtle ones) as a skilled chef who still made all his kitchen 'basics' by hand. (Black 1985, 10)

Given that canned food was not invented until early in the 1800s, this transformation is remarkable. The wholesale industrialization of the food system occurred hand-in-hand with the emergence of Fordist-style food companies like Kellogg's, Post, Coca-Cola and Pepsi. The transformation of food from unprocessed foodstuff to highly processed, highly marketed commodity is one foundation for the consumerism that is the hallmark of the current global economy (Walker 2004).

In this chapter we will explore the marketing side of this remarkable food-processing industry. Three aspects of commodification are germane to this investigation. First, as discussed in Chapters 1 and 2, the increased transportation and processing that ensued from the industrial revolution and urbanization, people were increasingly separated from where their food was grown. This meant eaters had to rely on a combination of food processors and regulators to assure them that the food they were consuming was safe. Second, the exodus from rural to urban centres was accompanied by decreased activity levels as people moved from physically demanding rural lifestyles to more sedentary urban ones. This meant that new food habits were needed as society adjusted to new economic and social realities. Granted factory work was not easy, but often times it was not as physically demanding as farming had been prior to the Industrial Revolution. Third, the intentional adulteration of food as explored in Chapter 2 led people to require further reassurance about their food. A common concern that emerged from these three aspects of the changing relationship between people and their food was the desire to eat safer food and a healthier diet. This quest began in the 1800s and continues to the present.

As early as the 1830s, Sylvester Graham a Presbyterian minister and founding father of the US vegetarian movement determined gluttony to be "the greatest of all causes of *evil*" (Stacey 1994, 13, emphasis added) and promoted the need for a vegetarian diet to maintain good physical and mental health, and curb alcoholism and sexual urges. In Graham's mind, food could be used to dampen sinful predispositions (Smith 2004). A supporter of the 'healthier' lifestyle promoted by Graham exclaimed,

> It often happens that what is plain and clear to the mental perception of a vegetarian, is obscure, if not wholly incomprehensible to the mind of a flesh-eater. (Stacey 1994, 31)

Hand in hand with food industrialization were shifts in the culture of how people related to their food. In a satirical response to food preoccupations of the late 1800s, it was noted that,

> Disease lurks behind the fat sirloin, and there is Death in the tureen of turtle soup. Whenever I go to a dinner party, it seems to me that I see in my mind's eye, the incarnate forms of Gout, Apoplexy, and Fever, bringing in the dishes and coaxing their victims, just to take one slice more. (Stacey 1994, 36)

Building on the thinking of Graham and fellow vegetarians, the Kelloggs brothers took up the cause of reforming eating habits as a way to promote a healthier, but also more productive America. In the mid–1800s, as part of the growing Seventh Day Adventist community in Battle Creek Michigan, the Kellogg family established a world-renowned sanitarium to promote vegetarianism, water cures and a healthy lifestyle. Abstinence was a cornerstone of the 'San' where the use of alcohol, tobacco, tea and coffee was prohibited. These New England Puritanical sentiments characterized the Kelloggs' childhoods,

> Lingering from the frontier days was some vestige of the severe, fun-denying precepts of the Puritans. Life still meant to most people, even children, hard work and long hours – the harder and longer the more praiseworthy. (Powell 1956, 25)

The 'San' attracted the cream of US society and included members of the Rockefeller and Penney families, as well as John Burroughs, Henry Ford, Harvey Firestone and S.S. Kresge as its patients (Powell 1956, 83).

The appeal of the 'San' lay in the Kellogg grasp of the need to make changes to American eating habits. A traditional breakfast for pioneers consisted of fried eggs, meat, fried potatoes, coffee, bread with molasses, and sometimes a slice of pie. This diet was not suited to a more sedentary, urban lifestyle,

> Very active people could handle such generous rations without too ill an effect. However, by the late nineteenth century, the struggle for existence had advanced beyond the pioneering and physical state. More and more people were engaged in sedentary pursuits. Machinery was taking some of the sinew-grinding characteristics out of labour. Yet eating habits, much like those of the pioneer who plowed and cultivated and chopped and reaped from before dawn until sunset, continued. Here and there in the nation farseeing individuals were taking stock of the harm resulting from improper diet. (Powell 1956, 85–86)

As part of the visionary group who realized the need for changes in eating habits, the Kellogg family developed a flaked cereal to replace meat at breakfast. By the 1890s the precursor to Kellogg's Corn Flakes as a standardized, safe and healthy processed food product was born. This sets the stage for the chain of food and diet issues that turn increasingly to increased production volumes and the application of more technology as a way to resolve food system challenges.

The Rise of Mass Marketed Food

Mr. W.K. Kellogg and the Kellogg cereal company are a shining example of the development and evolution of the industrial Fordist-style food system. Through the late 1800s and into the beginning of the 20th century, W.K. Kellogg applied assembly line concepts to food packaging and used mass marketing to promote his breakfast products. Mr. Kellogg had excellent modern business instincts and was able to propel the company forward by capitalizing on several opportunities. For example, following a devastating fire in 1903, Kellogg chose to site his new cereal plant close to both the Grand Trunk and Michigan Central Railway,

> Thus the comparatively new company president prepared to activate a cherished dream of modern factory buildings humming with activity twenty-four hours a day, with switch engines busily backing in freight cars of raw materials and strings of "empties" and hauling away loaded cars for many destinations. (Powell 1956, 118)

Kellogg had the vision to see the importance of transporting raw material into his plant and shipping finished product out. He saw his company as part of a much larger food system. Modernity and progress was the goal, as Kellogg himself stated, "I don't like this hand packaging. I want products that run off conveyor belts into packages" (Powell 1956, 130).

Kellogg was not alone in grasping the importance of his new business model. Forty-two cereal companies opened in the same county as the Kellogg factory between 1902 and 1904 as people rushed to cash in on the exploding demand for cereal flakes (Powell 1956). C.W. Post, a former resident at the San and an early supporter of the cereal production bandwagon, used a sanitarium recipe to launch his first breakfast beverage in 1894 under the name of Postum. With help from a Chicago ad agency, the tag line 'Makes red blood' was used to market the drink that was promoted as, "a builder of nerves, red blood, and all-around health" (Powell 1956, 104).

During the initial cereal boom, all was not tranquil in the Kellogg family ranks. The Kellogg brothers parted ways over disagreements about the power of advertising and what W.K. Kellogg saw as missed opportunities. The Kellogg brothers worked together for over 20 years building the San. The older brother – a physician and idea man behind the San products – did not want to use advertising as he worried that it would sully the reputation of the health institute. Mr. W.K. Kellogg, the younger brother, had no such fears. He understood that the public needed symbols they could identify and that would give them the confidence with regard to product quality and consistency. By 1906, the W.K. Kellogg signature appeared on every box of corn flakes as an assurance to customers that they were buying the real thing – a corn flake that was flavoured with malt, sugar and salt. The addition of sugar to the recipe was another point of contention between the brothers. Their disagreement foreshadows current debates in the conventional food industry around consumer health versus market share. Mr. Kellogg added cane sugar to the ingredients in 1905 despite Dr. Kellogg's strenuous objections about the potential negative effects on diabetics (Powell 1956, 110). Yet Mr. Kellogg was more concerned with sales, taste and securing market share for his company than he was with the overall health of the consumer,

> Mr. Kellogg realized that he now had a breakfast food superior to anything yet developed. He had seen competitors walk off with what could have been a sanitarium monopoly on cereal coffee, a granulated cereal and wheat flakes. Unless bold steps were then taken to capitalize on this new opportunity, the same thing could happen to corn flakes. He was about ready to take these bold steps. (Powell 1956, 111)

Within the space of one generation, the priority for the family business shifted from health to profit, and from nutritious product development to marketing. The promotion of Corn Flakes included the door-to-door distribution of samples, billboards, and streetcar and window signage. These were accompanied by personal salesmen visits to department stores and food jobbers. Kellogg was the first to use a city-focused blitz. In one of the first of these campaigns in New York City, Kellogg used a 'Wink' campaign that consisted of three stages. First, Kellogg used billboards and print media to advertise their 'Wink' Wednesday. Second, was the use of newspapers to encourage readers to "Give the grocer a wink, and see what you get" (Powell 1956, 135). Third, grocers were given supporting product and advertising material that included the following instructions,

> Here is one of the "Wink" Ads that will appear in all of the leading New York papers, both morning and evening on Wednesday, June 5th.
>
> Thousands of feet of billboard will also be in place by that time.
>
> This advertising will arouse the curiosity of the entire city. It is bound to bring the people into your store. Your customers will either wink or ask the meaning of the wink.
>
> When they do, give them one of the sample packages of Kellogg's Toasted Corn Flakes herewith. That is what these samples are for – they're the secret of the wink.
>
> With this package of samples you'll find a poster. Please hang this in or on your window Wednesday. It will form the connecting link between our advertising and your store. (Powell 1956, 135)

Sales of Corn Flakes to New Yorkers rose from under two carloads per month prior to the ad to one carload a day after the ad.

Another innovative strategy developed by Kellogg was to link a personality to his product. In 1907, Kellog's launched the 'Sweetheart of Corn' – the Corn Flakes poster girl. Kellogg's also realized the value of high quality print ad and engaged the services of American artists including J. C. Leyendecker and Norman Rockwell. Kellogg also appreciated the impact of size and hype. By 1912, Kellogg had a 106 x 50 foot lit advertising sign in Times Square and a moving electric sign in Chicago. (Powell 1956, 139) The goal was to make the product a trusted household name as reflected in ad copy when an attractive young woman exclaimed, "Excuse me – I know what I want, and I want what I asked for – TOASTED CORN FLAKES

– Good day" (Powell 1956, 140). The product was also promoted as nutritious and satisfying,

> In two jiffies a flavoury meal to satisfy the hungriest man – Kellogg's Crispy Corn Flakes.

> 1st It's a meal to make a hungry man beam with joy. With milk or cream, as satisfying as it is tasty.

> 2nd Good for every man who uses up a lot of body-fuel. Builds energy fast.

> 3rd. It is ready to eat. No cooking. No waiting. Just pour in a bowl and serve. (Powell 1956, 140)

This ad captures the essence of food advertising – trust, satisfaction and convenience. The ad reflects the expectation that women of the day would want to 'satisfy their hungry man'.

As the decades went by, and media options changed, the focus on marketing continued, and Kelloggs showed that it also understood the benefits of marketing to children as they pitched television ads during circus shows and other programs for youngsters. The company sponsored radio and television shows , so for example, "of great appeal to children in early radio days was the 'Kellogg's Singing Lady' who interspersed fairy tales with songs" (Powell 1956, 139). We will return to the topic of advertising and children later in the chapter when children are targeted more specifically due to their direct and indirect influence over billions of dollars of spending per year. Between 1905 and 1940, Kellogg spent $100 million on advertising – and grew his products into household names that people trusted. The marketing of processed food was so successful that by 1940 Americans consumed almost as much processed as fresh food, with the level of processed food increasing by another half by 1954 (Walker 2004, 248–249).

During the post WWII period in US history, consumer spending and income also changed. By 1955 Americans controlled five times the discretionary income they had at their disposal fifteen years earlier. This presented an interesting challenge to manufacturers. According to Packard's 1957 book, *The Hidden Persuaders*,

> "…one big and intimidating obstacle confronting stimulators [advertisers] was the fact that most Americans already possessed perfectly usable stoves, cars, TV sets, clothes, etc." Without the impulsive spending of those discretionary funds, warehouses would overflow and the economy would grind to a halt. The solution was for advertising to equate old products with feelings of inadequacy or inferiority, such that "by the mid-fifties", as Packard relates, "merchandisers of many different products were being urged by psychological counsellors to become 'merchants of discontent'". (Dale 2005, 36)

And, so we see signs of the societal shift to accepting – either tacitly or knowingly – planned obsolescence of consumer goods and over-consumption of consumables.

The food industry in its current form was built on convenience and product variety. The number of food products introduced to the public starting in the post WWII era spiralled as food processors sought to add value through processing to food. In other words, a company makes much more money selling frozen broccoli with cheese sauce than it does selling a fresh head of broccoli. Campbell's soups are a case in point. Campbell's and its flagship product Campbell's canned tomato soup were embraced by consumers. The product was seen as modern and progressive – a sanitary, easily stored, and convenient product – symbolic, in fact, of the best of the postwar, mass-production modern society. Pop-artist Andy Warhol was made famous by capturing these daily objects of American Fordist mass production, and his 1964 Campbell's Soup Can silkscreen is one of his most famous works of art.

A dominant concept – perhaps the dominant concept – in the food industry for decades has been to differentiate within the market place by creating new products. For example, in 1998 over 11,000 new 'food' products were launched in the United States – 37% of these products were categorized as candy, gum, snacks or condiments (Nestle 2004, 25). Even the desire for healthful, convenient food manifests itself in a myriad of apparently contradictory products. Jolly Time Candy Apple Flavoured popcorn is a good example of these inconsistencies. This microwavable snack product promotes itself as a fat free healthy option –one that contains artificial sweeteners and is artificially candy-apple flavoured. The company president explained in a press release for the product launch that, "JOLLY TIME is banking on the concept that *more choice equals more sales* from the growing number of health-and weight-conscious consumers" (emphasis added, Jolly Time 2004). In this case, corporate food processors capitalize on anxieties of people's poor health to carve up the food market into ever-smaller niches. This strategy is not unique, of course, as evidenced by the rapidly growing line of products that are "trans-fat" and "sugar free" (Bostrom 2005).

Retail Capitalization on Fear

By the 1980s, the Kellogg Company returned to its roots as it began to develop health-focused products that fed the American baby-boomer desire for immortality. In 1984 they launched All-Bran cereal as an anti-cancer weapon. In so doing, they reflected back the growing public desire to control their bodies, possibly beat death, and, at the very least, manage risk (Stacey 1994). This period seems to signal a new preoccupation with issues related to body space and the incorporation of food as an 'outside' force within (Whatmore 2002; Goodman 1999). Industrial food companies use science and the manipulation of nature to profit from this preoccupation to give consumers what they want. Part of their success lies in the distance between consumers and their food, and the perceived need to control and conquer nature as the means to quell fear and regain some control in an unpredictable and chaotic world. Dave Nichol, of Canadian-based President's Choice fame, understood these fears and used retail and brand marketing to capitalize on them.

Dave Nichol is considered by many in the food retail industry as the master of understanding the importance of brands to consumers. As President of Loblaw International Merchant (LIM), a subsidiary of the international Weston dynasty, Nichol translated this understanding into increased profits and carved out a name for himself in food retail history. Nichol is the quintessential 'foodie' who travelled the world in search of the ultimate food experiences. This quest exposed him to a range of retailing strategies including the St. Michael's product line at M&S and Trader Joe's in the United States. Before delving further into Nichol's role in the food industry, it is important to explore some of the models he drew from in developing his strategies.

M&S is an original innovator in retail branding. The chain of stores was established in 1894 while the exclusive St. Michael's trademark was registered in 1928. The St. Michael's line of products started with clothing, other textile and household products. By 1946 the retailer was engaged in a technological innovation overhaul to make suppliers more competitive in the post WWII era and, "to assist manufacturers in the progressive modernization of their plant and to adapt themselves to the latest technical advances" (Briggs 1988, 126). In 1948 the retailer added food to their St. Michael's line as part of the move to product differentiation and upgrading. A consistent priority for the product line – that has continued to the present – is domestic sourcing, high quality and good value. Although the store offers a limited range of products, the company dictates product lines to its manufacturers and has a tight vertical integration for its products.

Emerging from WWII, M&S entrenched itself in the British retail landscape by providing a successful example of home-grown retailing. In the depressed economic environment following the war, M&S was a much-needed source of pride for the country so that its success, "was a national asset whose growth was a matter of satisfaction to everyone" (Rees 1969, 193). Increasingly, food was an important part of the profit margin so that by 1968 food accounted for 25% of sales. Standards for food stressed freshness, cleanliness and taste. British farmers were encouraged to use superior inputs and efficient, industrial processing and distribution techniques – what de Somogyi referred to as a 'technology approach to marketing' (de Somogyi 1967, 56). Hygiene was also a top priority and set new standards in areas of public health serving as a model for UK hospitals of the day (Rees 1969). By 1985, all 260 stores carried only the St. Michael's brand that was over 90% British made (Tse 1985).

Yet the line of products was disconnected from fashion and food trends in the 1990s that privileged global over local (Rose 2007). A foray into the international retail market in the 1990s was largely a failure and the stores reconsolidated in the UK (Burt et al. 2002). Entering the 21st century, the shedding of unpopular 'dowdy' lines (Rose 2007, 53) and a stronger focus on food, re-established M&S as a force to be reckoned with in the UK retail industry. According to M&S CEO Rose,

> We took a calculated gamble on heavily increasing our food advertising budget in the months immediately after I took over. The first new ad we did was simple: We showed the beautiful food we offered. Internally, we called it food pornography, because it zoomed in on every morsel, with a sensual voice-over from a well-known actress. It was a hit. Sales of the chocolate pudding shown in the ad increased by 3,000% before Easter in 2005. It was a needed morale boost – and a signal to investors that we were on our way back up. (Rose 2007, 58)

More recently consumers have been demanding local, quality food. The attraction for the retailer in meeting this demand is the higher than normal profit margins as middlemen are removed or squeezed by the power of M&S. Although the emphasis on quality suits the retailer, onerous paperwork and exacting expectations make the relationship less desirable – though necessary – for farmers. It also demands more interaction with farmers as M&S put 'a face on farming'. As individual farms are identified with products, traceability is also a high priority. In store 'meet the farmer' events further impose on farmers' time (M&S 2007). Further, although farmers are guaranteed a stable market if they meet the standard, there is huge downward pressure on price. This usually translates into lower wages for farm workers as labour is the only variable cost farmers can manipulate. In some cases, farmers earn below their production costs. Clearly, this is not sustainable (Drottberger 2005). As food journalist Joanna Blythman put it, the buyer-seller relationship is "one-sided, largely feudal, and very unfavourable to the grower" (Drottberger 2005, 40). For consumers the new food focus means,

> ...[b]eing a 100% own brand label gives us a unique ability to control the quality of our food. We also have an unrivalled team of experts – Agronomists, Farmers, Chefs, Winemakers, Animal Welfare experts and even a Marine Technologist – to help source the best tasting foods which are produced to the highest standards. Our new range of speciality foods – from Smoked Wild Salmon from Alaska to Oisín Venison from the Finnebrogue estate and Greek Halkidiki olives from Mount Athos – all comes from people who truly care about their craft, whether it's our experts who source our food, or producers who are passionate about the ingredients they use or grow. (M&S 2007)

Recent research about the M&S Oakham chicken underscores the value of marketing quality food to UK consumers (Jackson et al. 2007). In their account of this new house brand of chicken, Jackson and his colleagues develop a very nuanced study of the use of 'alternative' by the high-street retailer. In this case M&S use the language of the 'alternative' movement to sell their newly branded chicken. While making the point that there are simultaneously vendors selling dubiously 'alternative' product at farmers' markets, Jackson et al. (2007) are quick to point out,

> ...the Oakham story is emblematic of a wider process occurring within the food industry whereby 'mainstream' retailers such as Marks & Spencer are appropriating the language that was formerly associated with 'alternative' producers. Through the development of the Oakham White brand, we argue, companies like Marks & Spencer are trying to get the best of both worlds, combining the reliability and quality assurance that is associated with intensive 'scientific' modes of production with the traceability and concern for tradition and taste that consumers are increasingly demanding (Jackson et al. 2007, 329).

Trader Joe's served as another template for Dave Nichol. Trader Joe's is a chain of privately owned grocery stores headquartered in Monrovia, California that works on the same model as M&S – value and quality – with the added twist of food sourced directly from producers and processors from around the world as they cater to new cosmopolitan, ethnic and exotic food preferences. The German-based Albrcht family trust owns the chain valued at $5 Billion.[1] The founder, Joe Coulombe started the chain in 1958 as Pronto Market. By the late 1960s, the winning format for the store was perfected and the Trader Joe's stores were launched. According to the Trader Joe's website the strategy is to,

> ...travel the world in search of interesting, unique, great-tasting foods and beverages. We buy direct from the producer whenever possible. We strip away all the fancy stuff and focus on the important things like natural ingredients and inspiring flavors. (Trader Joe's 2007)

In addition, the company emphasizes corporate frugality and boasts "no corporate jets or fancy offices" (Trader Joe's 2007). The result for the consumer is low prices for both staples and exotic food.

Keeping the successes of M&S and Trader Joe's in mind, we now return to Canada and Dave Nichol. Learning from the successes of M&S and Trader Joe's, Dave Nichol used his personal food experiences and his own attitudes about food to create food products that spoke to the ever-sophisticated demand of consumers in Canada's fast-growing dynamic city-spaces. According to biographer Anne Kingston, Dave Nichol inherited his food attitudes from his parents,

> To his father, food was sustenance; eating your share and no more was virtuous. His mother on the other hand, equated cooking with love. She valued abundance...Even as a child, the young Nichol [Dave] seems to have had an uncommon interest in the subject. (Kingston 1994, 14)

The contradictions between food as pleasure and penance recur throughout the Nichol story, and in many ways reflect North America's own torn relationship with food.

Nichol joined George Weston Ltd. in 1971. In 1970 the company posted sales of $1.4 billion making it the largest food company in Canada and third in the world (Kingston 1994). In the early 1970s the Canadian Loblaw operation – part of the Weston empire – needed an overhaul and Dave Nichol was part of the Loblaw executive restructuring team, that included Galen Weston, Richard Currie, designer Don Watt and Brian Davidson. The first step in the bid to revamp the company was to increase profitability. As part of these efforts, Nichol brought advertising in-house to improve control over production costs. The decision was made to challenge household name national food brands such as Coke and Kellogg for market share. The goal was to increase profits from the then standard 1.5 to 2% up to 5% by providing quality food at reasonable prices. The needed innovation to realize this goal was to have Loblaw supervise the manufacturing of its own product lines. Products were sourced from around the world and then packaged under the gourmet President's Choice (PC) label. Outsourcing (known as co-packing in the industry) kept control over taste and quality in Loblaw hands but meant that Loblaw did not have to invest in processing infrastructure. It also gave Nichol the ability to pit one

---

1 The family also owns the international ALDI grocery chain with over 5,000 stores throughout the EU, US and Australia (Aldi 2007).

processor against another and keep the Loblaw profit margin high. Finally, it provided leverage against branded food manufacturers as their shelf space was squeezed by PC product. This allowed Loblaw to build confidence in brands associated with their chain of food stores. As Kingston explains,

> The supermarket industry had been structured to reinforce and perpetuate the dominance of the trusted national and international brands – Procter & Gamble, Nabisco, Kraft, and General Foods. In the vortex of the supermarket, brand names served as a beacon. They signified a product's origin, like the brand given livestock by their owners or the thumbprint the first potter gave to their wares. The word itself derives from an Old Norse word, *brandr*, which means to burn. In modern times, the brand name of manufactured goods has assumed a larger symbolic meaning. It is a talisman, a pact between producers and consumers. Producers implicitly vow that the product will taste or perform in a consistent way. Shoppers accept that promise with their purchase. (Kingston 1994, 63)

As Dale elaborates,

> Brand loyalty had long been a central concept in American marketing, and many familiar brands introduced as far back as the 1880s (Uncle Ben's Rice and Aunt Jemima pancakes) had been designed to offset the sense of social loss that occurred when packaged goods replaced the old-time shopkeeper, who used to scoop staples like oats and flour from bins for his customers. (Brand names, in other words, were surrogate personalities created for an increasingly depersonalized world.) (Dale 2005, 47)

But Nichol was after more than one product people could trust. He sought to build a relationship between a line of branded products that ranged from 'PC Italian Dog Food' to 'Memories of Kobe 2 Minute Marinade' to 'G.R.E.E.N.' toilet paper. The goal was to,

> …[o]ffer solutions to the modern anxiety that looked to food as status, as entertainment, as consolation, as redemption. (Kingston 1994, 294)

Nichol used his quarterly publication the *Insider's Report* – modelled and named after the Trader Joe in-house newspaper[2] – television, print and in-store ads to hype a product. The *Insider's Report* was central to a product launch as Nichol used it to create product credibility and cachet.

The quality and pitch of a product was critical to Nichol's success and reflects his ability to read the consumer. As a result, the evolution of PC brands is a history of shifts in consumer preferences and preoccupations. Beginning with more traditional products such as gourmet coffee in the mid 1980s, the trend changed to ethnic cuisine and included products such as PC Extra Virgin Olive Oil and a line of 'Memories of…' sauces including Sonoma Dried Tomato Sauce, Jaipur Curry and Passion Fruit Sauce, Montego Bay Jerk Marinade, and Fuji Shitaake Mushroom Sauce. Nichol understood the need for up-market, quality food that had an exotic, global flare and catered to the demand for "processed ethnicity" (Kingston 1994, 154). Nichol also

---

2 The name *Insider's Report* was purchased from Trader Joe's in 1983 and the Trader Joe's paper was renamed *The Fearless Flyer* (Kingston 1994, 47–48).

grasped the opportunity to build trust with the consumer through brand products and the profit premium that branded products could add to the bottom line.

To address 1980s consumer environmental preoccupations with issues such as the hole in the ozone layer, Nichol launched the G.R.E.E.N. label. This product line focused on the consumers' need to feel they were making a contribution towards environmental remediation. Chlorine free diapers and paper products and phosphate free laundry detergents are the core of this line. In this case, the PC brand was used to forge trust between the retailer/food manufacturer and the consumer in a world of growing uncertainty.

By the late 1980s and early 1990s health concerns re-emerged as the aging boomers confronted their mortality and sought ways to improve their quality of life. As the preoccupation with environmental issues in the early 1980s gave way to concerns about personal health and food quality, Nichol changed gears to develop products in synch with anxieties to reap the benefits of the change as, "[p]ublic obsessions had shifted from political to the personal, from altruism to self-interest, from ozone levels to cholesterol levels" (Kingston 1994, 216). The renewed interest in healthful eating provoked Nichol and his colleagues to launch a line called 'Too Good to be True!' that included diet products, 'Ancient Grains' cereal, and 'Beta Blast Beta Carotene Cocktail'. At the same time, they offered a line of products that promoted self-indulgence to the boomers. This line, with the Decadent Chocolate Chip cookie as its star, capitalized on the desire to indulge and repent,

> ...[a]mong foodies who measured morality in terms of what they ingested, the word itself [referring to 'Decadent' the name of the President's Choice chocolate chip cookie] had come to signify a forbidden pleasure. To a fitness obsessed population, indulging in fattening food was a form of sin, a temporary lapse into moral depravity. Penance was paid at the nautilus machine. (Kingston 1994, 120)

Nichol and the team were masters when it came to understanding shifts in consumer demand. The conflicts inherent in these product lines were lost on the consumer as the *Too Good to Be True* line offered both "sin and then redemption" (Kingston 1994, 243) – it was the perfect way to increase market share and profit margins for the food retailer while reassuring the consumer. The *Insider's Report* and PC brand foods had another role in an increasingly angst-ridden society where,

> North Americans had lost the art of impersonal civility, without which urban life could be nasty and brutish. People no longer spoke to strangers, not even about the weather; the rift between those who lived in climate controlled condos and those who spent their winters sleeping on street grates was growing wider. But the Insider's Report also played to a darker fear – the social and political insecurity that increasingly found its focus in food. It exploited uncertainty about the safety of drinking water and mass manufactured food. (Kingston 1994, 73)

With this brief overview of retail marketing in mind, it is now interesting to explore the effects of advertising on consumer buying. In the next section we examine the relationship between advertising, food-related diseases, and the effects of advertising on children.

Marketing Safety or Complacency?

In a Cornell study funded by the United States National Institute for Commodity Promotion and Economic Evaluation and the National Science Foundation (Messer et al. 2006) research was conducted to determine the effects of advertising on consumer buying patterns. This study was interested in determining the influence ads could have on purchasing patterns for food that has been compromised in some way. The particular focus of this research was on the relationship between a consumer's understanding about Mad Cow disease, advertising and hamburger consumption. Participants were divided into four groups: Group A saw a video explaining the presence of Mad Cow disease in the US beef system and the fatal results of contracting the human form of Mad Cow; Group B was shown a promotional video from the National Cattlemen's Beef Association; Group C saw both videos; and Group D, the control group, did not watch any videos. Adult participants were then asked how much they would be willing to pay for a freshly cooked hamburger. Group A indicated they would pay $0.88, Group B said $2.52, Group C $2.07 and Group D $2.14. In this case, all people were still willing to buy the burger despite the potential fatal association with BSE, while the difference for people who saw both videos and the group that saw no videos was 3.5% (or $0.07),

> These results suggest that advertising may play a valuable role in helping inoculate consumers from dramatically changing their purchasing behaviour in the face of reports of uncertain health risks associated with various foods. (Messer et al. 2006)

It is clear from this research that advertising has the potential to radically impact public perception about food borne illness,

> It is interesting to note the historic response by the beef industry in the wake of the December 2003 discovery of the first cow infected with mad cow disease in the United States. Less than a month after the discovery, the National Cattlemen's Beef Association quickened the release of a new advertising effort and increased funding by an additional $1.3 million.

> Perhaps in part due to this action, subsequent beef industry research suggested that the percentage of consumers who were confident that US beef was safe from mad cow disease did not decrease. Thus, from the beef industry's perspective, their decision to increase advertising at that time might have been an optimal response. (Messer et al. 2006, 1)

Targeting Children: The Ultimate Frontier

Advertising is part of the modern childhood landscape. As the National Academy of Science explained,

> Among the various environmental influences, none has more rapidly assumed a central socializing role for young people than the media, in its multiple forms. With its growth in variety and penetration has come a concomitant growth in the promotion of branded food and beverage products in the marketplace. (McGinnis et al. 2006, 2–4)

Ads have become even more of a force in children's lives since the 1980s when advertisers realized the influence children have on parental spending and their potential as life-long consumers (Schlosser 2002). This is especially relevant for the food and beverage industry. There is strong evidence TV ads influence both the food and beverage preferences and purchases of children between 2 and 11 (McGinnis et al. 2006, 8). Marion Nestle, in her book *Food Politics* (2004) explains the connections between the increased power of kids as consumers of foods high in fat, salt and sugar (HFSS) and advertising. In 1997 kids in the US spent over $60 billion on food, most of it sugary, empty calories. Children aged 7–12 spent the most money on candy with over 50% of their spending in this category. Next in line were chewing gum, soft drinks, ice cream, salty snacks, fast food and cookies (Nestle 2003, 177–178). Children also wield influence over food purchases made by their families providing input into 25% of salty snack purchases, 30% of soft drinks, 40% of frozen pizza, 50% of cold cereals, and 60% of canned pasta (Nestle 2003, 178). In an effort to maintain or grow these numbers, ads targeted at children promote HFSS foods. The highly processed nature and low input costs of these foods means that processors and retailers can make a healthy profit selling these products – well above the margins for other food products. Soft drink manufacturers including Coca-Cola, Cott beverages and Cadbury-Schweppes, reported net profits of 9.4% in 2007 (Britvic 2006). In the UK, Britvic is self-described as, "one of two leading soft drinks businesses in Great Britain both by volume and retail sales value...Britvic has had successful relations with Pepsico since 1987, which was renewed in 2004 for a further 15 years". Britvic when listed for the first time on the London Stock Exchange in December 2005 was valued at £10 billion. Britvic reported a net profit for the year ending October 1 2006 of 39.6% with pretax profits of 55.9% (Britvic 2007). This compares with net profits of 5.9% for the major diversified food processing companies including industry giants such as Kraft, Unilever, and Archer Daniel Midland.[3]

The investment by food processing corporations in TV advertising reflects the faith placed in this form of promotion and the desire to promote sugary foods and beverages. For example, in the UK the Office of Communications reported in 2003 food, soft drinks and Chain Restaurants spent £522m on TV ads. This equaled 72% of their budget in this category. In 2005 Coca-Cola, Pepsico and Britvic spent £56 million on advertising (Britvic 2006). Food processors dedicate extensive resources to advertising as processed food represents a substantial portion of consumer spending. In 1999 it accounted for 12.5% of consumer spending in the US. Of equal importance is the fact that over 80% of US food products are branded so ads are needed to keep products in the public eye (Gallo 1999 in Story and French 2004). The goal of advertising directed, "especially [to] young children, appears to be

---

3 According to *The Economist* (2007) the top ten food, beverage and tobacco companies reported the following sales and profits: (in billions) Nestle: 76.11; 5.89 Altria Group 68.92; 10.44 Unilever 49.35; 4.95 Archer Daniels 35.83; 1.02 PepsiCo 32.56; 4.08 Tyson Foods 26.01; 0.35 Bunge 24.28; 0.53 Coca-Cola 23.1; 4.87 British American Tobacco 20.66; 2.11 Sara Lee 19.13; 0.55

driven largely by the desire to develop and build brand awareness/recognition, brand preference and brand loyalty" (Story and French 2004).

Despite claims by some skeptics that ads do not have an effect on target audiences (for a discussion of the pros ad cons of this point of view refer to Dale 2005), advertising investments appear to reach the target market. Of the average 17 hours of weekly television viewing by UK children, 154 minutes (or nearly 11%) are spent viewing children focused ads (Ofcom 2004, 16). There is little debate as to whether ads influence children's behaviour, but there is a fierce discussion concerning the age that a child is able to distinguish ads as separate from television programming. The age range varies from six to twelve depending on who you listen to, but the bottom line is that not all children can understand the difference between their shows and the ads (Dale 2005, 68). In the late 1990s child psychologist Reinhold Bergler conducted a study in Germany with over 1600 participants age 6 to 13 for the European Commission to examine the effects of advertising on children (Bergler 1999). When exploring a child's ability to understand that advertising is intended to sell something, Bergler discovered that 42.9% of six year olds, 14.6% of ten year olds and 11.2% of 12–13 year olds were unable to appreciate the role of advertising. In more recent research, Ofcom reported that the average eight-year old child understands that ads are intended to persuade them about a product and by eleven most children have developed a critical grasp of television ads (Ofcom 2004, 17). These findings are significant given that 26% of US children under two are reported to have televisions in their bedrooms (National Institute on the Media and Family 2005).

There is also a quantity consideration to this media phenomenon because the amount of advertising a consumer is exposed to affects the assimilation rate for products. So, for example, in a study of teen buying it was found that exposure to brands increases brand purchasing (Greenberg and Brand 1993). So, for kids who watch television, more TV equals an increased desire for the toys and food they see advertised (Strasburger and Wilson 2002). Other relevant facts brought to light in a recent paper by Story and French (2004) include: 75% of US food manufacturer ad budgets and 95% of fast food budgets are spent on TV ads; US kids may have seen 360,000 TV ads by the time they graduate from high school; 55% of TV ads in Saturday morning programming were for HFSS foods or fast food restaurants. Finally, a Kaiser Family Foundation study released in 2007 found that half of television advertising time for youngsters is for food, and of the ads surveyed (over 8,800) no ads promoted fruits or vegetables while only 15% promoted healthy lifestyles. They also found that children in the 8-12 year category watch the most TV (Kaiser Family Foundation 2007). All of this is not surprising given that children have yet to form a mature understanding about healthy food, are at the threshold of brand attachment, and that the largest profit margins are available in processed foods.

Concern about the profound impact of food advertising is justified when one considers converging circumstances so that: 1) the period of childhood when children begin to have their own money to spend is, 2) also the period during which they are forming their own food habits, and, 3) children are simultaneously unclear about the difference between the reality of advertising and the fantasy of television.

Television, of course, is not the only medium used to put food products in front of children. Product placements, in-school marketing, kids' clubs, promotions and the Internet are all used to flog food to children. In-school contracts have attracted much attention in recent years. For example, Channel One in the United States is a 12 minute broadcast (2 minutes of ads and 10 minutes of current events) that is shown in over 10,000 middle and high schools in the US (Story and French 2004). The company *Cover Concepts* approaches in-school advertising from a different perspective. As they explain on their web site,

> Cover Concepts, a division of Marvel Entertainment, has been providing FREE materials to over half the nation's schools since 1989. Cover Concepts works in tandem with administrators and teachers to distribute sponsored materials such as book covers, educational comics, teacher's guides, posters, bookmarks, and specialty packs to name a few.
>
> As America's largest resource for FREE classroom materials, Cover Concepts reaches 30 million kids in grades K-12, 1.2 million kids in daycare centers, 5 million kids in libraries and 750,000 kids in summer camps nationwide. (Cover Concepts 2003).

Cover Concepts advertising clients include McDonald's, Pepsi, Gatorade, Frito Lay, General Mills, Hershey, Keebler, Kellogg's, M&Ms, Mars, Kraft/Nabisco, Wrigley and State Fair Corn Dogs. Cover Concepts,

> ...[d]istributes textbook covers, lesson plans, posters, bookmarks, sampling programs, specialty packs, and lunch menu posters to participating companies. These products are branded with the company's name or corporate logo and then distributed free to students and schools. (Story and French 2004, 6–7)

The Internet is another emerging opportunity for food advertising. As of March 2007, just over 67 (67.8%) of Canada's population was connected to the Internet, while the US and UK reported 69.9% and 62.3% respectively (Internet World Stats 2007). There are countless ways companies are connecting with children online. Some of the tools used are branded corporate web sites with music, games, contests, downloadable screen savers, and newsletters. The Kellogg's Apple Jacks web site is typical. The site features interactive, animated games with characters associated with the cereal (Kellogg's 2007) so kids can now spend hours playing games that have clear product connections. Food companies also link themselves to children's entertainment websites such as Nickelodeon, Disney and Fox (Story and French 2007).

One of the most renowned brand marketing success stories is McDonald's – the success is so widespread that the term McDonaldization was coined to describe the penetration of the restaurant into the cultural and economic fiber of countries around the world (Ritzer 2006). Since its beginnings in the 1950s, the McDonald chain of restaurants used mascots – the most successful was Ronald McDonald – to brand the restaurant and create a connection with children in keeping with the example set by the leviathan Walt Disney Corporation. Schlosser describes the rationale for the 1980s increase in advertising to children in his best-selling book Fast Food Nation (2002),

...[h]oping that nostalgic childhood memories of a brand will lead to a lifetime of purchases, companies now plan "cradle-to-grave" advertising strategies. They have come to believe what Ray Kroc and Walt Disney realized a long time ago – a person's "brand loyalty" may begin as early as the age of two. Indeed, market research has found that children often recognize a brand logo before they can recognize their own name. (Schlosser 2002, 43)

Eating under the Golden Arches becomes an entertainment experience where "fantasyland...playgrounds and free toys" overshadow the super-sized food servings and poor nutritional value of the food (Dale 2005, 56).

Advertising directed at children is very sophisticated. It appeals to children in two ways. First, because children are consumers with their own money to spend, it appeals to them directly through issues relevant to their age group. Second, as gateways to the family bank account, advertising provides kids with an appropriate toolkit to nag their parents – what McNeal at Texas A&M University called "nagging tactics" (McNeal 2002). 'Nagging tactics' use pleading, persistence, tantrums, threats and pity to convince parents to buy a product. Advertisers provide kids with words to use when asking for a product and a vocabulary compatible with the views of their parents. Patriotism, national defense and health are all a part of the advertising lingo fed to children (Schlosser 2002).

Although advertisers and some experts paint kids as sophisticated consumers who are simply being given product information to use in their own best interests (Dale 2005, Rushkoff 1999), jurisdictions around the world have expressed concern or taken action to protect children against what are perceived to be the ill effects of advertising. In 1978 the US moved to regulate advertising to children through the Federal Trade Commission (FTC). The initiative came about as there was increasing concern about the amount of sugary food being widely marketed to children. After reviewing 60,000 pages of submissions, the FTC decided to leave the onus on the industry and 'terminated' the creation of rules in 1981 (Engle 2004). According to Marion Nestle (2006) it is time to reconsider this position. Nestle cites the Institute of Medicine report (IOM) released in 2006 which expresses grave concern about advertising to children,

> Food marketing, the IOM says, intentionally targets children who are too young to distinguish advertising from truth and induces them to eat high-calorie, low nutrient (but highly profitable) 'junk' foods; companies succeed so well in this effort that 'business-as usual' cannot be allowed to continue. (Nestle 2006, 2528)

Nestle urges the US to take decisive action consistent with steps in other countries. For example, Sweden banned all advertising directed at children under 12 in 1991. The same policy applies in Norway. Food ads targeted at children under 14 are banned in Australia and the Netherlands banned candy ads to the under 12 group (Nestle 2006). A Quebec ban on food advertising to children under 13 has been in place since 1978 (Fidelman 2006). The UK will bring their latest non-broadcast advertising guidelines into effect in July, 2007. This revised set of restrictions is meant to address Internet and other non-TV ads. However, the proposal has disappointed consumer, teacher and health groups who saw this as a chance to, "Finally make some progress, but they have fallen well short of the mark" (BBC 2007). The critics are asking that the new regulations include provisions for the frequency and duration, and that there are distinctions made between healthy and unhealthy food. Finally in early 2008 the Toronto Board of Health called for a ban on HFSS food ads to children under thirteen in an attempt to reduce future childhood obesity rates.

Food Marketing Complexities

This chapter lays out specific contradictions and tensions between food retail, processing and advertising goals and the well-being of consumers. In the case of the retailers the key to profit is to some extent about playing on consumer fears and uncertainty. In North America the relationship between food and consumers is extremely complex. In some cases food is about sin; in others, it is about satisfaction and gratification. Food marketers do their best to ensure they lay the groundwork for complex relationships between children and food as early as possible. Powerful advertising and marketing efforts – which cost money – are focused where volume and profit can be built. So as profitable niches are identified and developed, large companies swoop in and fill the gap – Wal-Mart organics is a clear example of this approach in action. Other examples are discussed in the chapters to come.

Having reviewed salient features of food regulation, retail and marketing, in the next chapter we are interested in moving to the production end of the food chain to find out how farmers fare in the modern food system. In determining the extent of trust placed in different members of society, farmers were highly trusted in the US (73%) versus the UK (43%) (Gaskell et al. 2003; Priest et al. 2003). Context is important here and could provide a warning for North American farmers. By 2003 the UK had suffered through the Mad Cow crises where a good deal of the blame was laid squarely at the feet of the farming community. By comparison, US farmers were held in the highest esteem when compared to other groups (the other groups included in the research are the media, industry, ethics committees, consumer organizations, environmental groups, government, shops, farms, churches and doctors). This social capital should not be squandered. It appears that some farmers may appreciate this trust and seek to protect it through increased surveillance and rigour. In the next chapter we explore the pressures exerted on farmers by the industrial food system and how different farmer groups are responding to the forces of globalization.

Chapter 4

Growing Distances: The Separation of Farmers, Ecologies and Eaters

> Realized net income for Canadian farmers fell for the second consecutive year in 2006 to its lowest level since 2003. Rising interest, wage and fuel costs, together with falling hog receipts and program payments, more than offset increases in revenue from crops and cattle. (Statistics Canada, 2007)

This chapter explores the links between the overwhelming economic pressures exerted on farmers by the industrial food system, shifting production practices and consumer food fears in Canada and the United States. As we understood in Chapter 2, there are strong parallels between Canadian and US policies towards agriculture. As we shall see in this chapter, these commonalities have resulted in similar economic pressures for farmers. In Chapters 2 and 3 we used political economy theory to understand how capital accumulated and regulations developed in support of IFS. This chapter builds from these observations, and also draws from political ecology theory. Political ecology is useful as it provides a context in which to make connections between policy, power and local ecological concerns. It helps us to understand the levels of complexity that have been constructed through the IFS. Although political ecology has not been used widely in a developed world context, McCarthy (2002) challenges researchers to apply this perspective to North America. In illustrating the relevance of political ecology for first world countries, McCarthy describes the themes that tie together political ecology as,

> Access to and control over resources; marginality; integration of scales of analysis; the effects of integration into international markets; the centrality of livelihood issues; ...the importance of local histories, meanings, culture, and the 'micropolitics' in resource use; the disenfranchisement of legitimate local users and uses; [and] the effects of limited state capacity. (McCarthy 2002, 1283)

As becomes evident from the following case studies, farming in North America is subject to international market pressures, power struggles and the loss of local community capacities. All of these tensions converge to threaten local ecologies. Both cases – the first about organic farmers taking Monsanto and Bayer to court over GMO seeds, and the second about two farmer associations and their different visions and actions with respect to ecology and local food production/consumption connections – reveal fundamentally different approaches to farming. On the one hand, the majority of farmers are disassociated from the ecology of their land through their engagement with the IFS and its products such as GMO seeds. This has the effect of making farmers party to issues that engender consumer uncertainty by growing food from seeds that some consumers mistrust, and by not being able to farm in a way that connects them to either their land or their local communities. On the other hand, there is a small network of farmers who maneuver within the constraints imposed by industrial food regulations to create alternative spaces for production-consumption relationships and construct healthier relations with their seed and the land (Schroeder 2005; Peet and Watts 2004). Analysis of ecologically focused farmers reveal that they are embedded in the local community and focus on balancing economic, ecologic, and community goals.

This research draws on work by McCarthy (2002) who compares the 'Wise Use' movement in the rural American west that existed in resistance to the dominant, non-local discourses of land management in that area. McCarthy's work explored the two oppositional views and the way they inform and then construct discourses about local ecologies. The research reported in this chapter shares this method as it explores disparate discourses about local ecologies. The case is made that the ecologically sensitive farmers[1] circumscribe the economic quagmire created by industrial agriculture to develop and realize their own imaginings about the shape and substance for a new food system (Haenn 2002, 4). By constructing a vision that includes ecological considerations, they insert themselves into the gaps in the existing system and are able to capture new producer-consumer dynamics. In this evolving relational production-consumption network farmers reconnect with, preserve and rebuild local human, cultural and land ecologies. The result is two parallel farming systems – one embedded in local communities that demonstrate concern for farm ecologies and another that focuses on export market development with few ties to land management or local community.

Exploring local ecologies is instructive and broadens the scope of our discussion as it engages us with the theme of sustainability. To date, this idea has not been fully incorporated into food studies and it needs more attention as a potentially seminal part of the way we envision new food systems (Maxey 2007; Ilbery and Maye 2005; Buller and Morris 2004). Maxey makes the point that this needs to remedied as sustainability provides a multi-scaled starting point for communities and jurisdictions to envision balanced goals (Maxey 2007). Pointing to work by Bowler (2002), Maxey reiterates the value of the triple bottom line that is the foundation of sustainability as all factors need to be considered 'simultaneously' and that while, "this sounds remarkably simple, it represents a paradigm shift from dominant Western reductionism and contemporary neoliberal privileging of the economic". (Maxey 2007, 60; see also Morgan et al. 2006)

To set the context, we begin with the production challenges farmers face in Canada and the United States. After that we examine the vertical integration that typifies the supply side of the IFS. With this context established we then examine the two case studies in Canada. In the first case we explore the Saskatchewan Organic Directorate (SOD) and its dispute with Monsanto and Bayer over GMO seeds. The

---

1 Here ecological farmers use organic growing methods to the extent possible and can be third-party certified or not. They emphasize on-farm biodiversity, healthy soil, air and water and aim to reduce fossil fuel use (NFU 2005a).

second case study presents two farmer associations in eastern Ontario and the way they are each responding to global market pressures.

Food Production in Canada and the US

When we consider the amount of land used for agriculture in the United States – 41.5% – and the number people employed in agriculture – less than 1% – we get a sense of the overwhelming amount of industrialization on the farm (Statistics Canada 2007a).[2] Obviously, this was not always the case. For example, at the turn of the century, even after the effects of the early part of the industrial revolution, 74.4% of people in Iowa lived in rural areas (State Library of Iowa 2005) while in Canada in 1901, 63% of people lived in rural communities (Statistics Canada 2005). This shift is a key element to the changes in agriculture over the last few decades and has shaped the way farming as a lifestyle and occupation has developed. With fewer people living on farms and able to advocate on behalf of agricultural issues, the importance of family farming has slipped off the policy table.

Despite attempts to improve efficiency through increased consolidation and mechanization as a way to manage shrinking profits, the economic picture is bleak for the few people left in farming. The statistics on Canadian and US food production underscore the extreme challenges that face farming communities and reinforce what farmers have reported anecdotally – that years of adjustments to global pressures and federal agricultural policies have hollowed out the ability of the average producer to farm for a living. On average, Canadian farmers[3] earn 17.7% of their family income as net farm income (Table 4.1).

Even in the category of farmers where the family farm is classified as business-focused and grosses over $250,000 annually, farm income still only accounts for 39.5% of net family income. This is not a short-term dilemma for farmers but part of a multi-decadal trend that has deepened since the 1980s. The decline has occurred in tandem with federal agricultural policy in Canada and the United States that encouraged and supported farmers in their efforts to get big and embrace export market focused agriculture.[4] Governments in Canada over the last two decades, as in other developed and developing countries, have encouraged farmers to consolidate land holdings, increase mechanization and apply more chemical inputs, as producers are actively encouraged to chase export markets through policy and programmes.

The uncertain farm economy has had another effect. Despite overall population growth in Canada, the number of farms decreased to just less than 230,000 by 2006, a drop of 7.1% from the 2001 census (Statistics Canada 2007b). This decline is consistent across the country and over the long term. In Ontario the number of farms fell from nearly 200,000 in 1921 to 57,211 by 2006. There was a decline of over

---

2 In the United States farming is no longer an occupation registered through the census as less than 1% of the population declare themselves as farmers.

3 The figures reported in this section are for unincorporated farms. Unincorporated farms accounted for 187,770 out of a total 246,923 farms in Canada in 2001, or 76% of all Canadian farms (Statistics Canada 2001; Statistics Canada 2004).

4 For example, the 1968 USDA advice to 'get big or get out'.

Table 4.1   Average total family income and sources of family income for farm families on unincorporated farms, Canada (Statistics Canada, 2001a)

Typology	% of all farm families	Average total family income in 2000	Percentage of family income from:					
			Net farm income	Wages and salaries	Self-employment income	Income from government	Investment income	Other
Retirement	20.3	54,520	15.8	22.7	2.5	29.8	13.3	15.9
Lifestyle	22.6	93,177	2.1	82.9	4.6	3.4	3.9	3.1
Low-Income	8.8	16,228	1.6	49.9	5.0	32.7	6.5	4.3
Business-focused up to $249,999	39.6	58,612	28.2	53.0	4.0	8.2	4.1	2.5
Business focused $250,000 and over	8.7	68,265	39.5	43.5	3.8	6.6	5.0	1.6
Total	100.0	62,695	17.7	56.7	3.9	10.8	5.8	5.1

11% in the number of Ontario farms between 1996 and 2001, and a further decline of 4.2% between 2001 and 2006 (Statistics Canada 2007c; Blay-Palmer et al. 2006).

And while the number of farms decrease, farm size increases. Between 2001 and 2006 farm size grew from an average 676 to 728 acres, an average increase of 7.7% (Statistics Canada 2007c). But despite the scaling up in farm size, the income benefits to farmers of increasing economies of scale supposed to be associated with export agriculture are questionable. Canadian farm revenue between 1979 and 2004 compared to export market activity for agriculture over the same period shows agricultural exports increased while farm income was stagnant or in decline. Supported by federal government policies such as NAFTA, estimated agricultural export values in 1979 were $5 billion and climbed to over $25 billion by 2003. On the other hand, net agricultural revenues for the period stagnated below the $5 billion mark, ranging between approximately $1 billion (1983 and 2002) and $4.5 billion (1981, 1989 and 1996) (Pellerin 2006). So while net export values increased, profits stagnated or fell.

The same is true for farmers in the United States. Despite tens of billions of dollars in government farm subsidies, US farmers cannot make ends meet. In his study of farmers in southern Minnesota, Ken Meter (2006) describes a typical US farm region and the reality of falling farm incomes. In 2002, the region reported 8997 farm families with a collective total of over 1 million farm animals. For the seven years to 2003, this part of Minnesota reported accumulated losses of $587 million. Meter attributes farm production losses to annual average deficits of $84 million as production costs exceeded farm income. As in Canada, farmers make up this shortfall partly through government support, but also from off-farm work. The situation is similar for the entire state of Minnesota where, overall, farmers lost $7 billion over the same 7-year period and in Iowa where farmers lost $4.78 billion.

The demographic shift that attends this economic decline is illustrative: while farm income falls, the average age of farmers is increasing. The data demonstrate very clearly that there are not enough young people farming and that a crisis is looming. The median age for Canadian farmers as of the 2006 agriculture census was 52 years, an increase from the median 47 years old reported in 1996 (for Ontario the median age in 2001 was 50). Of equal concern is the connection between the aging farm population and the declining number of new farmers. Between 1991 and 2001 the number of Ontario farmers less than 35 years old decreased by 7.7% while 25% of Ontario farmers collected a Canada pension (vanDonkersgoed 2006). In the US the picture is even more troubling as the average age of farmers in 1997 was 54.3 (USDA-ERS 2002a).

Vertical integration is another pressure exerted on farmers (Walker 2004; Friedmann 1993; Winson 1993). While different degrees of pressure are exerted on dairy, beef, pork and poultry farmers by the corporate food production chain, chicken production is one of the most vertically integrated examples in farming today (Hinrichs 2003). While we will consider the health implications of factory farming in Chapter 5, we focus here on the pressures exerted on producers from corporate integration. Eric Schlosser (2002) describes the vertical consolidation of Tyson Foods and the effect on chicken farmers in his documentary-style book *Fast Food Nation*. In order to supply McDonald's with chicken McNuggets, Tyson developed a new breed of chicken. This chicken, with unusually large breasts, was introduced in 1983 and within one month McDonald's became the nation's second largest US chicken buyer (after KFC). Tyson is currently the world's largest chicken processor. It achieved this status using factory principles now typical in the food industry. The company and its divisions are vertically integrated so that different parts of the Tyson conglomerate breed, slaughter and process chicken while it contracts out the raising (and the risk) to independent farmers. Tyson delivers the chicks they hatch at 1 day old to the farm. The chicks live their entire seven week lives on the grower's property but are regulated by Tyson who supplies feed, vet services and technical support, dictates feeding schedules, equipment upgrades, and does checks to ensure rules are being followed. Tyson maintains constant control over all aspects of 'production'. At the end of the seven weeks, the company picks up the birds, transports them, counts and weighs them. Although the grower is paid by the number of live birds produced and their weight, they have no way to verify the records produced by Tyson as the birds are collected on the farm and counted at the sight of slaughter – any birds that die along the way are at the 'growers' expense. In this scenario, the farmer provides the fuel, land, labour and chicken houses and assumes all of the risk. Financially, growers are beholden to Tyson as a house (of 25,000 chickens each) costs about $150,000 US and growers can't get a bank loan without a purchasing contract. And they cannot sell their chickens without agreeing to Tyson's terms.

This dynamic between farmers and corporation is not new. According to Schlosser, a 1995 Louisiana Tech Study found contract poultry growers earned about $12,000 annually – perilously close to the poverty line. And, the consolidation continues. The top three food product companies in the world – Archer Daniels Midland, Tyson, and Bunge – collectively reported over $86billion in revenues and more than 160,000 employees for 2005.[5] These companies progressively expanded their operations to include control over all aspects of the food production chain (McGirt 2006). For example, Bunge operates around the world, with its processing focused primarily in South and North America. As the largest oilseed processor in the world, their operations begin with the purchase of oilseeds and from there include: oil crushing facilities to produce food for animals, aquatics and humans; transportation networks to collect raw material and distribute product around the globe; and the processing and manufacturing of food items for sale to the food services market, other food processors and consumers (Bunge 2007).

If we combine the loss of control over family farms due to vertical integration with declining farm income, the decreasing number of farms and the aging farm population, two conclusions emerge. First, in order to keep their farms and grow food, the average farmer subsidizes the agri-food system with off-farm income work.[6] The secondary effect of this necessary commitment is that the average North

---

5 The level of corporate concentration is not unique to the food industry. According to a Fortune report in 2006, the combined revenues of the top 500 companies accounted for one third of the world's GDP (McGirt 2006).

6 This has occurred despite the importance of the food sector to the economy. For example in 2003 the agri-food industry provided over 700,000 jobs to the Ontario economy (11% of

American farmer is left with little time to farm as they spend more time working off-farm to make ends meet. Second, as farming is only a partial income-source, farming has become a rural life-style choice as opposed to a reliable income stream. As a result farmers turn to time saving measures and technologies as they balance farm, economic, ecologic and family considerations. One outcome is the scaling up in farm size and increasing amounts of mechanization such as the factory farms typical of Tyson.

While there are more sustainable ways to farm, the lines between conventional and ecological farming are fuzzy, as the following example of Earthbound Farm Organic demonstrates (Guthman 2004a, b). Earthbound produces and processes millions of servings of salad every year that are shipped across North America daily. Earthbound grows and processes its salads in accordance with USDA Organic standards and HACCP handling guidelines. Their salads products are grown on over 34,000 acres by 150 farmers, who farm anywhere from 5 to 680 acres. Earthbound places a premium on its environmental protection accomplishments, listing on their web site that through organic farming techniques it will,

> avoid [the] use of over 370,000 pounds of toxic and persistent pesticides;
>
> avoid [the] use of nearly 11.5 million pounds of synthetic fertilizers;
>
> [have] conserved an estimated 1.89 million gallons of petroleum by avoiding the use of petroleum-based pesticides and fertilizers; and
>
> fight global warming by absorbing carbon dioxide, a greenhouse gas, at an estimated rate of 3,670 pounds per acre, which for us is the equivalent of taking more than 7,500 cars off the road. (Earthbound Farms 2007)

In the wake of recent loose-leaf spinach scares in 2006 when spinach was found to be contaminated with E. coli, Earthbound is particularly attentive to food safety achievements in their promotional material. They explain the constant, thorough and rigorous testing used on their salad greens. This surveillance begins with their seeds and ends with product delivery. For example, their seeds are all tested for E. coli before planting. Once their salads are ready to be harvested, they are cold chilled within an hour of being cut. They use high tech sorters, chilling and washing facilities to process the greens. As described on their web site,

> Our custom-designed equipment inspects, washes, and dries our delicate salad greens without damaging them.
>
> Greens are sorted as they enter the wash line to remove weeds and undesirable leaves.
>
> Non-leafy items like roots and twigs are ejected by state-of-the-art optical sorters.
>
> The greens are thoroughly washed and sanitized in chilled, chlorinated water meeting National Organic Program standards before they're packaged. (Earthbound 2007)

Ontario's employment) while food and beverage products created $21 billion in food store sales (OFA 2004).

Detractors from Earthbound criticize the operations for the continental distribution of their operation, the elaborate packaging of their products and their monoculture style growing practices. The blurred lines between the conventional and alternative will be elaborated again in Chapter 7 as we examine the case of the SunOpta company in Canada. For the present, the important take away point is that as both alternative and conventional operations get larger, issues of sustainability can become less of a priority.

Having elaborated some of the challenges created by scaling up production within both conventional and organic farm industries, the rest of the chapter explores the different food systems that can emerge as some farmers are able to envision food systems in new/old ways. In some cases, farmers buy into the 'get big or get out' mentality, in other cases farmers advocate for more sustainable, community-centric farming practices. What we uncover are farmer activists, confused consumers, and a North American farming community under siege. As in other countries, genetically engineered food emerges as one of the sites of contention (Morgan et al. 2006; Los 2006; McAfee 2004; Whatmore 2002).

The Case Studies

SOD and Genetically Engineered Food

Genetically engineered (GE) seeds have been controversial since their introduction into the commercial domain in 1996 (McAfee 2004; Evenson and Gollin 2003; Whatmore 2002). Since then the debate has raged over important issues including: the impact of GE seeds on small-scale farmers in developing countries and their rights to the intellectual property developed over many generations and embedded in the seeds (e.g. Shiva); the health implications of GE food for consumers including allergens; trade issues between the EU and other developed countries; and not least, the ethics of manipulating the genetic code of plants. Although compelling, these topics are beyond the scope of this book. Instead we will focus in this section on GE technology and Canadian prairie farmers. In particular we will review the attraction of GE crops for farmers, the claims made by MNC seed and chemical companies, and the response of some farmers to the introduction of GE seeds. These issues are of interest as they fuel uncertainty about the food system for some consumers, consequently adding to food fears.

First it is useful to understand GE technologies in the context of the separation of nature and society as discussed in Chapter 1. In the case of GE of food crops, the proponents of GE technology use a free market perspective to frame discussions about genetic material. In this discourse, genetic material is presented as interchangeable pieces in plant genomes as though genetic bits are cogs in a machine and not organic material from a living organism that may be ingested into the human body (McAfee 2003, 2004; Whatmore 2002). This approach ignores the role of genetics and the broader dynamics between ecologies, biologies, economies and cultures,

The values of nature are equated with the prices, in actual or hypothetical international markets, of natural resources such as timber and medicinal-plant samples and of ecosystem services such as tourism sites, CO^2 sequestration, and water filtration. This approach is reductionist in that it treats nature and its components as quantifiable and as separable, at least conceptually, from their contexts in living nature and society, while it obscures the effects of political, cultural, and ecological factors on market transactions and resource values. (McAfee 2003, 204)

Using this line of thinking GE companies have taken the next logical step and sell GE seeds to farmers as merely one more 'weapon' in their arsenals against agricultural pests and declining farm profits. Building on the themes from post WWII when pesticides and herbicides were described in scientific terms, and then layering on the more combative, controlling language of the 1970s to 1990s (Kroma and Flora 2003), packages of chemicals and GE seeds are now marketed as a team of products that combine to defeat weeds and improve yield. Chemical products such as DuPont's product Battalion™ is said to, 'Keep you [the farmer] in command'. North American farmers use BASF products named 'Arsenal' and 'Stalker' to keep weeds under control. And Syngenta describes its 'Boundary' herbicide in a 2001 ad as providing protection against weeds so that, 'Annual grasses, broadleaf weeds and nightshade just can't get in when new Boundary is on guard' (Better Farming 2001a). The Aventis ad for the herbicide 'Converge' appeals to the same 'take no prisoners' mindset, "IT KILLS WEEDS – THEN COMES BACK TO GET THEIR FRIENDS" (Better Farming 2001b). In order to increase the sale of its GE seeds, Monsanto encourages farmers to combine GE with advanced chemical technologies so they, "Upgrade to YieldGard Plus with RoundUp Ready Corn 2" for improved pest and weed control, convenience and higher yields (Monsanto 2007).

In the effort to reduce farming to a controllable business model, farmers are treated by MNCs as the front line in the manufacturing of food. Farmers are the producers of raw materials for the IFS where production quantity using the most sophisticated equipments and technologies is the primary goal. Although the majority of North American farmers, in reaction to the scale and cost pressures exerted in the vertically-integrated food system, subscribe to this approach as the most effective and efficient way to farm, others opt for different farming strategies. Organic farming is one example of an ecologically focused farming approach. As a result of trying to preserve their alternatives, the second group of farmers can find itself pitted against seed and chemical companies (Holloway et al. 2005).

The story of the Saskatchewan Organic Directorate (SOD) and their fight to keep GE wheat off the Canadian prairies is one example of this type of fight. SOD – a group of organic prairie farmers and their supporters – faced many challenges as they fought to conserve production options for the future, preserve relationships with their communities and farm according to ecological principles. The struggle began in 2002 when two Saskatchewan farmers launched a court case against Monsanto Canada Inc. and Bayer CropScience Inc. The farmers claimed they had suffered losses on their organic farms due to the introduction of GE canola onto the Canadian prairie. The claimed losses included the inability to grow GE-free canola due to genetic drift from GE canola plants. They also sought Class Action status so that all organic farmers in Saskatchewan could be represented – and potentially compensated – from the court action (SOD 2007a). The Saskatchewan Court of Appeal handed down the most recent decision on May 2, 2007. The ruling was important for several reasons. First, it denied the farmers' application for Class Action status. The Court of Appeal also upheld earlier rulings that shifted the responsibility away from Monsanto and Bayer and onto organic certification bodies and the Canadian government. The certification bodies were deemed to be at fault as they required organic products to be GE-free as part of the certification standards between two and three years after the introduction of GE canola. The court also held the Canadian government responsible as they had approved the varieties for sale. According to the ruling,

> As it pertained to canola, the action had a number of weaknesses. The Government of Canada had approved the unconfined release of Roundup Ready [the Monsanto product] and Liberty Link [the Bayer CropScience product] canola well before the action was commenced. And the private organizations in the business of certifying organic grain farmers did not have any standards in place regarding the presence of genetically modified organisms until well after Roundup Ready and Liberty Link canola had been made available and become widely accepted. Only then did these organizations amend their standards to preclude the presence of genetically modified organisms in grain marked as "organically grown". Moreover, the companies do not grow canola but only make their varieties of seed available to farmers wishing to do so. (Court of Appeal of Saskatchewan 2007, 3)

Although the farmers were not successful in gaining their Class Action status or in obtaining compensation for their losses as organic canola farmers, they did succeed in halting the introduction of GE wheat into the prairies; in 2002 Monsanto decided to cease the process to introduce GE wheat in Canada. However, as SOD cautions, the removal of GE wheat by Monsanto offers no guarantees for the future, and 'vigilance' is needed to ensure that more GE crops are not approved in Canada (SOD 2007a). In addition to the legal action, SOD maintains an active presence on behalf of Canadian organic farmers and provides input into matters of relevance. In an on-going effort to prevent the introduction of other GE crops, SOD wrote a position paper opposing the introduction of GE alfalfa (SOD 2006) citing the uncontrollable spread of alfalfa pollen and the inevitable contamination of non-GE alfalfa. In 2007 SOD submitted another paper commenting on proposed changes to Seed Variety Registration in Canada. Their concerns are that, "[t]The proposed regulation threatens quality, access, public accountability, and the buyers' [farmers'] right to unbiased information about seed" (SOD 2007b, 1).

As this brief case study of SOD makes clear, there are many tensions that pull farmers in conflicting directions. First, the farmer is the target of MNCs as potential buyers of GE seeds and chemicals. Second, according to the courts, the Canadian government has done a poor job regulating and overseeing the introduction of GE seeds into Canada. But as noted in Chapter 2, there are strong bonds between government and corporations that undermine the ability of the government to act as watchdog on behalf of the public. And, the focus on the bottom line by MNCs – such as Monsanto and Bayer – and their influence on governments meant that in this case, at least, the farmer was left to defend not only their own rights to future growing options, but to look out for the public good as well.

The abdication by the government of its role as overseer of public wellbeing undermines trust in the food production system. Some farmers, such as those belonging to SOD, step into the breach and assume an overseer role. However, the role of farmers as watchdogs does not meet the full range of consumer needs. This is borne out by work from Rutgers University that explores consumer awareness about GE content in the US food supply and the level of public trust in the food system in this context (Lang and Hallman 2005; Hallman et al. 2003). Despite a high desire to have food labeled as containing GE ingredients (94% of Americans), US food is not labeled accordingly. In addition, awareness about food and GE ingredients is remarkably low. Only half of US consumers were aware that their food contains GE ingredients, while only 25% of Americans know they have ingested food with GE ingredients[7] (Hallman et al. 2003). The same disconnect exists in Canada. In a 2004 report prepared by Agriculture Canada, it was reported that 77% of Canadians have concerns about the safety of food that contains GMOs. In research polls about food safety, consistently 85% or more Canadians want food containing GMOs to be labeled as such (e.g. DECIMA 2004, Environics 2001). But despite a clear message from the public and experts about their concerns (Royal Society of Canada 2001), the federal government not only resists regulating, controlling, and labeling, but continues to pour tens of millions of public research dollars every year into food related biotechnology research (Blay-Palmer 2007, 14). Regardless of the concerns and demands from the public[8] food with GE ingredients is not labeled in Canada or the United States. The lack of response by government to consumer concerns and demands about GE food ingredients, the need for government protection of the well-being of citizens and their food, and the associated institutional indifference on the part of government and corporations for consumer concerns decreases the trust consumers have in the food system making them understandably wary about the food they eat.

In the context of this institutional void, we now turn to a research project in Ontario, Canada. This project offers insights into the mismatch that can exist between local economic development projects, the actual production potential of a region, and the need to develop relevant project goals. It also helps us to understand the pressures faced by farmers as they try to make their farms balanced viable operations. As part of this analysis, we compare the stewardship capacity of two farming groups – the Ontario Federation of Agriculture with the National Farmers Union. The study presents interesting insights about the two groups and the role their farmers assume as stewards of the land.

---

7 It is estimated that between 60% and 70% of processed food contains some GE ingredients (Byrne 2007).

8 In Canada a bill to label food with GE content has been voted down three times by the federal House of Commons.

Eastern Ontario Farming: Building Economically Sound and Ecologically Sensitive Farm Communities

The study area included two counties (Frontenac and Lennox-Addington) and the City of Kingston, all located in eastern Ontario. Kingston is approximately halfway between Montreal and Toronto on Lake Ontario. With a population of over 115,000 (Statistics Canada 2001b), Kingston has an economy grounded largely in the public service sector. It depends heavily on universities, colleges and medical and correctional facilities for local employment. Queen's University in particular is central to R&D projects in the city (KEDCO 2004). The surrounding rural counties are sparsely populated with pockets of small towns scattered throughout the study area. The research project reported here took place over two years.

Understanding actual versus imagined production capacity   The project goal was to identify existing local farm-based economic activities that could be reinforced and new opportunities that could be developed. As the project goals were being established, enhancing local fruit and vegetable production and related processing activities such as preserving and freezing were imagined to be realistic economic development opportunities. Encouragingly, when surveyed over 80% of local shoppers and all the retail buyers we interviewed expressed a strong interest in buying more local produce. However, when the actual production levels for produce were examined it became clear that in the short-term at least, production capacity was very low. A comparison of regional production data between 1951 and 2001 showed that vegetable and fruit growing had decreased in all categories by up to 95%, with an average decrease of 84%. The total acreage for vegetables fell from 2623 to 449 over the 50-year period. The only exception to this decrease was strawberries, which saw an increase of 55 acres. So although the historical data indicate the potential for increased fruit and vegetable production and enhanced local self-sufficiency and sustainability, in the present, farmers have moved away from labour and time-intensive fruit and vegetable production to more mechanized farming.

The decline in acreage dedicated to produce returns us to the discussion about decreased farm income earlier in the chapter. As we know, if farmers cannot make a living from farming they need to do off-farm work. As well, in many cases farmers choose to increase the level of on-farm mechanization to increase efficiency. Market gardening does not fit into this model of farming. There is however a core group of ecologically focused farmers in the study area who are exploring market garden and direct sell options. We will learn more about them later in the chapter. For now we will focus on the majority of farmers in the region and the opportunity to raise beef for the local market.

How to rebuild a local beef system   The preliminary findings guided the research to questions about actual production activity as a starting point for short-term development. We used Statistics Canada (2001) data to establish existing activity in the local agricultural sector. The analysis revealed the importance of dairy (14% of all farms in the study region) and beef farming (51% of farms had beef operations) supported by substantial acreages of hay, fodder and cash crops (over

155,000 acres in 2001). On the demand side, we estimated the study area consumes approximately 11.6 million pounds of beef annually. The region births on the order of 15,000 head of cattle annually. This is equivalent to an estimated 9 million pounds of beef. It follows then that if the beef supply chain were reorganized and the number of animals in the two counties stayed constant, it should be possible to consume all the beef produced locally. When one adds the fact that beef is responsible for 30% of local food miles there is a double dividend potential for increased local economic activity and decreased environmental impact by raising more animals for local consumption (Lam 2006). Given these numbers, the potential for local beef production-consumption became a primary focus for the project. And so we set out to understand the challenges facing the local beef industry.

As mentioned earlier, our research with consumers and retailers revealed strong support for buying local. Over 80% of the 200 consumers surveyed said they would buy local product if they could easily identify it and had more information about the benefits of buying local for farmers and the community. All of the retailers interviewed expressed a desire to buy local products. Retailers did express the need for stable distribution channels (i.e. they can't deal with individual farmers, so there is a need for an intermediary) and prices need to be reasonable.

In order to understand the issues for producers, we had conversations with farmers and industry experts. These discussions confirmed that since the 1994 North American Free Trade Agreement (NAFTA) most farmers in the region have become part of the North American flow of meat products. Practically, this means that farmers raise calves to between six and eight months and then ship them outside the region – often to the US – to a feedlot to be fattened for about one year. At about 18 months, the animal is slaughtered and returned to the area as processed, often boxed, meat. This has occurred in many parts of the country and has created an industrial continental beef industry that has hollowed out local animal feeding and value-added meat processing capacity (Blay-Palmer and Donald 2006).

To produce and process beef you need animals, some kind of feeding system, and abattoirs. As we identified through the Statistics Canada data analysis, we would need more animals if all of the beef needs were to be met locally. However, assuming we decide to start with existing number of animals in the region, the problem arises when we look for fattening capacity, as there are now only two feedlots with a total capacity to finish about 500 animals per year. As well, there is only one abattoir remaining to kill and cut the meat. Further, at the time of the report writing (Spring 2006) the lack of locally raised mature, animals meant the abattoir owner actually had to import beef from as far away as Argentina to cut for the local market. Clearly, existing infrastructure is insufficient if the goal is to rebuild local food linkages. In order to offer options to farmers, the research team put together a system for sustainable beef production that: 1) kept animals on farms through to the slaughter stage; and, 2) identified feasible feed and financing programs so farmers would not suffer undue hardship as they transitioned to a new local beef system.

After the presentation of the report findings, it was interesting to observe the reaction to the proposal for a revamped local beef industry. Despite the wake-up call farmers had received as the US border closed overnight in 2003 as a result of BSE crisis – we will explore this in more detail in Chapter 5 – most Kingston region beef farmers were still focused on supplying into the North American continental supply chain. Instead of looking at other options they were looking for ways to get the border reopened. Given the average age of farmers, this is not surprising. It is difficult to learn new ways of doing things (Morgan and Murdoch 2000). This is especially true when approaching retirement – as many of these farmers are – and the value of their key asset – the farm – as a going concern as a stable operational/ commercial regime is key to the family's prospects.

Generally, farmers who support vertically integrated food production belong to the Ontario Federation of Agriculture (OFA). This group represents over 90% of farmers in the case study region. However, there is also a handful of ecologically focused farmers who have recognized the potential offered by a re-localized food system and have a vision for improved local food production and direct farmer-consumer connections. In large part, these farmers belong to the National Farmers Union (NFU). The two groups are compared in the next section as a way to explore the realities of food systems and farm ecologies.

Institutions and ecologies in Ontario farming   There are multiple actors and institutions that help to define agriculture and local ecologies in the study area. Those relevant in this analysis are: the provincial Ontario Ministry of Agriculture and Rural Affairs (OMAFRA) and the federal Department of Agriculture and Agri-Food Canada (AAFC), the OFA and the NFU. Earlier in the chapter, we discussed the broad economic context facing farmers. We will now consider the farmers' ability to consider land stewardship issues in the context of: 1) existing agricultural institutions, and; 2) efforts to enhance local food links. This will provide the clearest links in the chapter of the relevance of the political ecology perspective for these issues.

Government Institutions

This section assesses provincial and federal approaches to and supports for the ecological stewardship of farms. First, we present the approach taken by the provincial Ontario Ministry of Agriculture, Food and Rural Affairs (OMAFRA) with a focus on the post-NAFTA period. Second, we present the Environmental Farm Plan as an example of a cooperative, self-examination tool for Ontario farmers. Third, we discuss federal ecological priorities with respect to farms. In particular we look at the most recent round of consultations about the national agricultural policy. If we consider the approach inherent to sustainability, where economic, social and ecological issues are all considered simultaneously and in balance with each other, what emerges from the analysis is that neither level of government places a very high priority on farm ecology issues. Ecological concerns are certainly not given the same weight as economic considerations.

OMAFRA provided excellent extension support services to Ontario farmers up until the mid-1990s. However, a shift began in 1997 as services were converted from extension and on-farm support to export and marketing expertise and development. The provincial ministry made a deliberate decision to adopt an export and commodity focus and shifted its resources accordingly. The change in direction by OMAFRA coincides with NAFTA and reflects the emergence of a greater emphasis on international free market economic activity and the increased importance of globalization. Within this larger policy framework, the 'Foodland Ontario' label and the 'Environmental Farm Plan' both described here provide interesting insights into government priorities for farm ecology and sustainability.

The Foodland Ontario label was initiated in 1977 (Ilbery et al. 2005). The label is used to identify any produce grown in the province. It is especially visible in summer and fall when local produce is most widely available although the burgeoning greenhouse industry has given the label new visibility on a year round basis. The Foodland Ontario label is used across a range of retail and farmers' market venues. It reflects the 'old guard' approach to agriculture when OMAFRA saw its mandate as supporting farmers in local markets. In its current form, the mandate of Foodland Ontario is, "to promote increased sales of Ontario produce and to improve market penetration" (Foodland Ontario 2007). The label has always had an economic focus and was not designed to address either ecological or social challenges in farming. Although ecological benefits may accrue from this initiative due to reduced food miles, these consequences are not intended.

The Ontario Environmental Farm Plan (EFP) on the other hand has a very strong environmental focus. The EFP was launched in 1993 as a joint effort of the Ontario Federation of Agriculture (OFA), the Christian Farmers Federation of Ontario, the Ontario Farm Animal Council and AGCare (Agricultural Groups Concerned About Resources and the Environment) to address on-farm stewardship. The program is delivered locally by the Ontario Soil and Crop Improvement Association (OSCIA) and is supported by technical expertise from OMAFRA and funding from Agriculture and Agri-Food Canada (AAFC). The goal of the program is to identify,

> key issues as involving water quality (and especially contamination by agricultural practices), soil quality, air quality, agricultural inputs and natural areas such as wetlands and woodlots. It is recommended that every Ontario farmer should develop and implement an environmental farm plan tailored to the needs of their own farm, addressing issues of farm productivity and profitability, aquifer degradation, surface water degradation, farm health and safety, and air pollution (Furman, 1997)…Participation by farmers in the EFP is voluntary and, if followed in full, involves a six-stage sequence from attendance at an introductory one-day workshop to the implementation of the Plan. (Robinson 2006, 862)

Upon completion of the workshop, farmers return to their farm to conduct their own evaluations. The uptake of this program provincially is about 25% (Robinson 2006). Farmers who participate in the program are characterized as:

> better educated farmers, involved in community organizations with prior awareness of the existence of an on-farm environmental need, presence on the farm of a woodlot and/or wetland and greater overall concern for the environment. (Smithers and Furman 2003 in Robinson 2006, 865)

Clearly the goal of this program is to improve on-farm environmental quality. Robinson (2006) documents the reasons farmers participated in the program. The specific rationales given included the improvement of: 1) soil quality (through for example conservation tillage), 2) water quality (in part through better well condition and well management) and the related issue of 3) agricultural waste storage. These three categories together accounted for 52.5% of all activities undertaken under the EFP. Although these categories do reflect better environmental stewardship, they are improvements related either to production or liability. In the case of soil quality, the use of conservation tillage does help to reduce soil compaction as the farmers do not till the soil as much, but it does not remediate damaged soil profiles in any substantial way.

In the case of better well and water management, there is a heightened sensitivity to water contamination and liability in the wake of the 2000 Walkerton, Ontario crisis. As reported in the inquiry report into the tragedy, "seven deaths and illness in over 2300 people, including children who may endure lasting effects, were caused by E. coli O 157:H7 and Campylobacter jejuni contamination of the town's water supply by manure-based run-off from farmland (O'Connor 2002)" (Robinson 2006, 866). So in the area of water quality, one reason to participate in the EFP is to mitigate potential liability related to E. coli contamination. Other categories that represent more fundamental environmental improvements – such as energy efficiency (4%), woodlands and wildlife (1.5%), wetlands and wildlife ponds (1%), and water efficiency (1%) – account for only 7.5% of the total activities undertaken by farmers in the EFP. So even in the context of a deliberate program to address environmental concerns, most farmers are not able to make substantial differences to on-farm soil health or other issues of ecological concern. We now turn to a consideration of federal policy.

At the national scale, Agriculture and Agri-Food Canada (AAFC) have consistently supported large-scale export agriculture through public policy since the late 1980s. The current iteration is reflected in policy documents that were circulated to the public for comment in the winter and spring of 2007. In preparation for the consultation process about federal agricultural policy, 15 documents are made available online. Of these, two were framing and contact information documents; 11 have an economic focus (including one related to innovation and another on skills development and training); one discusses issues of food safety; and one deals with questions related to the environment. To understand the importance of ecological and environmental issues for the government, a content analysis was undertaken for the framing document (AAFC 2006a). The content analysis revealed that the word 'environment' is used 20 times. Four of these uses refer to issues of environmental stewardship. The remaining 16 references to 'environment' use the word to refer to policy and economic conditions. The word ecology does not appear in the document. The pre-occupation of the federal government with economic issues is underscored in the document outlining general economic opportunities wherein the AAFC explains,

> New opportunities for agriculture and agri-food products are emerging in three distinct
> areas. Rapidly emerging developing economies like China and India are expected to

generate the greatest source of new demand for traditional exports. In markets of the members of the Organisation for Economic Co-operation and Development (OECD), future growth is expected to be limited to niche markets for value-added products. The bioeconomy and its diverse range of possible new bioproducts offer potential not only to improve upon current product attributes through biotechnology, but also to create new sources of income by tapping into non-traditional markets. (AAFC: 2006b)

The language of the position paper makes it clear that the emphasis for the AAFC is on export market opportunities and high technology and not on responsible environmental stewardship. Other documents address economic benefits, structural issues, macro conditions, policy, trends, economic well-being and markets. The lone document that addresses environmental considerations states the agricultural policy objectives to,

- Promote a competitive and profitable sector that takes advantage of opportunities, meets environmental objectives and improves financial performance;
- Seek to ensure the sector has access to the resources (land, water, air) that it requires;
- Support sector adaptation to the impacts of the changing environment. (AAFC 2006c: 6)

The environmental policy objectives are to:

- Reduce the impact of agriculture on the environment;
- Contribute to broader policy objectives related to the environment, including promoting the health of Canadians. (AAFC 2006c: 6)

In this context, the federal government sees itself as a *facilitator* for the agricultural industry with the environment as an external factor that must be managed to ensure maximum profit. As they explain,

...governments may have a role to play in facilitating the adoption of technologies which improve environmental outcomes, through sound regulation or to support the sector in developing business models that provide a fair return on investment in agri-environmental management. (AAFC 2006c: 6)

OFA and NFU farmers all operate within this provincial and federal political and institutional framework. The different responses to the institutional reality reflect divergent visions for and approaches to food, farming, land stewardship and nature.

The Ontario Federation of Agriculture (OFA)

The OFA is the provincial incarnation of the Canadian Federation of Agriculture. Within each province, counties have a federation to represent local farmers. The OFA represents over 90% of farmers in the study region. The provincial federation commitment to industrial agriculture is reflected on their web site and through two recent campaigns – 'Farmers Feed Cities', and their support of income stabilization programs aimed at balancing the devastating impact of export-focused commodity driven agriculture. The thrust of the 'Farmers Feed Cities' initiative is described as,

...an agriculture awareness campaign working with farmers across Ontario. The campaign is a reminder that agriculture and the rural economy are important to Ontario. Right now, farmers are facing an income crisis due to circumstances beyond their control. This vibrant industry needs immediate help to fight foreign government intervention. Without it, our future is uncertain. (OFA 2006a)

The goal of the campaign is to raise public awareness about the plight facing farmers as a way to leverage income stabilization support from the federal and provincial governments. There have been public rallies as farmers converged on the provincial and federal legislative buildings in their tractors, as well as the collection of signatures on petitions, letter writing campaigns and BBQs. Attempts are made to 'put a face on farming' by connecting farmers with local people through public events. Personal connections are less important though than the overall message – that commodity farmers need financial stability.

Under the present circumstances members of the OFA are both the victims of and now co-creators – in a Foucauldian sense – of the institutional framework that promotes the large scale, industrial farming that ultimately undermines the soil, water and genetic resources that are the foundations of sustainable farming (Robbins 2004; Wolf 2001). Having accepted the commodity-based export market agricultural system – and now as promoters of that system – the only way farmers can envision making a minimum standard of living is through income stabilization programs. By staying large, and focusing on volume production of one or two crops, there is little room in this model for substantive stewardship. This is supported by the way in which the Environmental Farm Plan was adopted as farmers focused on production and liability. Under these circumstances, farmers do not have the latitude to act in a more ecologically sustainable manner since it cannot be one of their primary concerns. They envision a future for themselves that embraces the 'get big or get out' approach to farming that is the foundation of the globalized industrial agriculture system. The recent rush to embrace corn farming – what Birger calls 'The great corn gold rush' – as an input for ethanol production underscores the immediate consideration of profit over land stewardship.

Conventional corn farming is one of the most punishing forms of farming for the land and ideally corn is rotated with less harmful crops such as soybeans. Corn planting is on the rise though as corn prices doubled from $2 to $4 per bushel between 2006 and 2007. North American farmers are so optimistic about this new opportunity that corn planting and harvesting equipment is sold out in some locations as more and more acres are planted in corn (Birger 2007). This trend does not bode well for the health of agricultural land.

The National Farmers Union

The NFU, Branch 319 is the regional arm of the international NFU farm group. The NFU vision for food and farming is centered on a sustainable farm community,

...the NFU works toward the development of economic and social policies that will maintain the family farm as the primary food-producing unit in Canada...The NFU believes agriculture should be economically, socially, and environmentally sustainable. Food production should

Table 4.2   Comparison of local, social, ecological and economic goals as a priority for organizations operating at different scales

Institution/ organization	Initiative	Description	Local focus	Ecologic	Social	Economic
OMAFRA (provincial)	Foodland Ontario	Label developed to promote, educate and allow identification of provincially grown produce to 'increase market penetration', since 1977.	√			√
OMAFRA, AAFC and farm groups	Environmental Farm Plan	Provincial program to provide technical support and some financial assistance to farmers to undertake ecologically sustainable initiatives, since 1993.	√	√		√
AAFC (national)	2006-2007 consultation process	On-going public meetings to solicit input into the development of federal agricultural policy.				√
OFA (provincial, regional)	Farmers Feed Cities campaign	"An agriculture awareness campaign working with farmers across Ontario. The campaign is a reminder that agriculture and the rural economy are important to Ontario. Right now, farmers are facing an income crisis due to circumstances beyond their control. This vibrant industry needs immediate help to fight foreign government intervention."			√	√
NFU, Branch 319 (international, regional)	Feast of Fields	Annual harvest event begun in 2004 including local farmers, chefs and citizens to celebrate and educate public about local, ecologically produced food and local farmers. Receives local economic development support.	√	√	√	√
	Local label	'Buy local' label developed to promote, educate and allow identification of local food. Receives local economic development support.	√	√	√	√
	Conference and education support	One year program begun in 2006 to develop lecture series and conference to educate people about need for and benefits of local food. Receives AAFC support.	√	√	√	√

lead to enriched soils, a more beautiful countryside, and jobs of non-farmers, thriving rural communities, and enriched natural ecosystems. (Reynolds 2004, 1)

Although this group represents only less than 5% of farmers in the study region, they have initiated three programs since 2004 to promote local food. These programs are all focused on the creation and promotion of local farmer-eater connections that the NFU flagged as critical for future food system sustainability. For example, in describing the Feast of Fields, an NFU sponsored event, Reynolds reports,

> Organizers [of the Feast of Fields] note they want to promote local and seasonal food because local food is fresher, tastier and healthier than imported food and its impact on the environment is significantly less. (Reynolds 2004, 1)

The NFU is a catalyst for change in the Canadian food system as its members point to weaknesses in the industrial food system and offer local, ecological solutions to these challenges. They have authored several reports including, 'The NFU Policy on Sustainable Agriculture' (NFU 2005a), 'The Farm Crisis: Its Causes and Solutions' (2005b), 'Solving the Farm Crisis: A Sixteen-Point Plan for Canadian Farm and Food Security' (2005c) as well as reports specific to regional challenges such as the Submission by the National Farmers Union Region 6 (Saskatchewan) on 'Selected Rural and Agricultural Issues to the Government of Saskatchewan' (NFU 2006). The focus is on the overall health of the food system,

> Sustainable agriculture must be based on a mutual understanding between farmers and non-farmers. Farmers have an obligation to provide safe basic foods and to steward the soil, water, and air. Non-farmers, in return, must support farmers through fair prices and programs which [sic] protect farmers from loss of income and unreasonable eviction from their farms.

> The governments of Canada have surrendered much control over agriculture to transnational corporations. Current government policy, in effect if not intent, is often no more than the promotion of these corporations' agendas. Unfortunately, the agendas of corporate chemical, fertilizer, processing, distribution, and retailing corporations conflict with the best interests of farmers, farm families, rural communities, as well as with those of consumers.

> Farmers, farm families, local communities and regions must regain control of food production. (NFU 2005a)

NFU farmers acknowledge the challenge inherent to the dominant agricultural paradigm, and offer ways to deal with the farm crisis in a manner that sustains local ecologies, communities and economies.

Since the research study on food relocalization was completed, they have had some success. An organic greenhouse continues to operate at capacity. A beef and market garden farm has expanded its operations so it is now also running an off-farm kitchen that makes meat pies and other prepared products. And the area's first CSA has a waiting list of customers. Although these are small gains, farms are successfully breaking new ground. In the fall of 2007, the Food Down the Road initiative hosted a conference to discuss ways to link institutional buyers with local meat producers. And, the abattoir is still in business.

In this chapter we have added another layer to our understanding about the level of complexity in the industrial food system. Table 4.2 helps to make explicit the differences in approaches to ecological, social and economic factors between provincial, federal, OFA and NFU initiatives. Clearly the NFU initiatives all offer a more integrated and balanced approach to farming. This is consistent with the NFU goals of creating a more sustainable food system.

The farmers in the OFA are relatively more hooked into the global industrial food system through their commitment to export-based agriculture. Despite appeals to Canadian consumers to support them as 'Farmers feed cities' implying a local connection between the Canadian rural hinterland and its cities, their focus is predominately on feeding into the global food system. As links in the export agriculture chain, they are just as likely to feed people in the US (in the case of beef) or Japan (in the case of soybeans) as people in the 'local' cities of Canada. Moreover – within this OFA group – there is little attempt to engage in a personal way with consumers to build one-on-one trust based relationships (Penker 2006; Murdoch et al. 2000). In ecologic terms, the focus is on production oriented changes so that OFA farmers tend to be less concerned with questions of farm ecology, in particular wildlife, woodland and wetland management and more concerned with production based solutions such as reduced tillage. As the focus of the OFA is on income they appeal to the government for basic minimum income levels. This is consistent with AAFC consultation documents on future federal agricultural policy. Interestingly, there are no serious attempts to secure compensation for stewardship initiatives as in the EU Common Agricultural Policy even though this would be a way to merge ecologic and economic goals.

So, in the context of the research presented here, although demand for local produce outstrips supply, OFA farmers are not inclined or able to learn new ways of farming, and are not willing/ able to shift their production to meet local market opportunities (Morgan and Murdoch, 2000). Rather, they look to federal and provincial governments for more support as they seek to remain viable. The present circumstances for this group of "mainstream" farmers requires that conventional economic priorities take precedence over ecological sustainability, as well as local well-being giving way to global pressures (Larner and LeHeron, 2002).

The NFU on the other hand actively promotes local linkages, ecological farming and land stewardship. They are deliberately connected with and committed to the local social, economic and ecological well being of their communities. They have a soundly defined ecological focus and they integrate this into all facets of their operations. They also have strong support from the national and international branches of their organization. This enables the NFU and its members to understand the 'power geometries' that frame the global-local agricultural context (Massey 1997). Unlike OFA members, NFU farmers have decided to envision and now create a new system of agriculture outside the dominant paradigm. OFA members, by contrast, have bought into the dominant export focused discourse to the extent that they self-regulate and perpetuate the very system that undermines their way of life.

What is particularly interesting is the emphasis by the NFU on local relationships. This is relevant in the context of the final section of this chapter that connects ideas of embeddedness, power and current levels of complexity in the industrial food system.

Connecting the Dots: Embeddedness, Food System Complexity and Fear

As we have seen in this chapter, food systems are shaped by goals that reflect contested visions and discourses about food production and consumption (Maye et al. 2007; Le Heron 2006; Morgan et al. 2006; DuPuis and Goodman 2005; Smithers et al. 2005; Hinrichs, 2003). Embeddedness and the related connections between society, nature and economies is a useful concept in this context. Embeddedness can be traced back to Polanyi's insight about the difficulty inherent in having nature and society subsumed under economic activity,

> The term 'embeddedness' expresses the idea that the economy is not autonomous as it must be in economic theory, but subordinated to politics, religion, and social relations... He [Polanyi] uses the concept to highlight how radical a break the classical economists made with previous thinkers. Instead of the historically normal pattern of subordinating the economy to society, their system of self-regulating markets required subordinating society to the logic of the market. (Block 2001, xxiv-xxvi)

Since Polanyi, other academics have used the concept of embeddedness to understand the dynamic between society and the economy in an attempt to bring society back into the analysis of economic activity (e.g. Evans 1995; Granovetter 1985). More recently the idea of 'local' as one facet of embeddedness in the food and rural studies literature, has been incorporated into AFS analysis to help understand the ordination and interrelationship of food production vis a vis local consumers, international markets, and regulatory regimes among the gamut of influences (Sonnino 2006; Morgan et al. 2006; Sonnino and Marsden 2006; Watts et al. 2005). As part of this discussion, the role of nature has been (re)introduced into musings about economy-society relations (Penker 2006; Sonnino and Marsden 2005; Kirwan 2004; Winter 2003; Murdoch et al. 2000). For example, Murdoch et al. (2000) link food quality with physical proximity, trust-based supply relationships and ecologically sustainable production. More recently, embeddedness is used explicitly to assess the connections between social, economic and ecological considerations from the perspective of multiple and inter-related scales.[9] This literature proposes a range of embeddedness dimensions that help to refine the way we understand the workings of local food systems. As a result, several scholars have been able to clarify the tensions between the local and global forces that simultaneously embed and dis-embed food systems (Morgan et al. 2006; Murdoch et al 2000). Sonnino and Marsden (2006) make an important contribution to this endeavour as they explain that,

---

9 This is consistent with efforts in economic geography to consider: the interaction between scales and the search for the local (Cox 1998); Swyngedouw's 'nested scales' (1997) that describe scales acting at the same time and in related ways; and, Bunnell and Coe's (2001) 'containers' that bound spaces of innovation.

... to assess the level and degree of horizontal embeddedness of alternative food networks – i.e. the extent to which they are socioculturally, economically, and environmentally embedded in their locality – it is crucial to investigate the strategies used by different actors to start, consolidate, and develop these innovations. Often based on negotiated and socially constructed notions of quality, such strategies speak about power relations within and between food chains. Investigating how power is distributed across the food chain and, more specifically, how actors involved in alternative food networks see their role in challenging and reshaping the agri-food system is an essential step for understanding the nature of these networks and their potential for new forms of rural development. (Sonnino and Marsden 2006, 194)

When we apply this type of analysis to the case study, several points become clear. As we examine the origin of OFA power, we find a food system defined by global politics and economic forces that are beyond the control of eastern Ontario farmers. Under these circumstances – that have been both accepted and reinforced by this group of actors in the Ontario farm community – the best strategy farmers have been able to develop is to ask for more financial support from the government. On the other hand, the food system of the NFU while informed by the parameters established by its international governing body, is firmly embedded in local communities, economies and ecologies.

The comparison of institutional and organizational priorities is revealing and provides insights into how embeddedness is manifest through food systems and approaches to local ecologies. Political ecology is useful as a way to understand how policy emanating from different scales and power structures influence attention to local ecology. The NFU is embedded strongly within the local, while the OFA is focused primarily on export, commodity driven food production. The OFA takes its cues from the international domain as translated through the federal and provincial governments. The NFU, while cognizant of the pressures from the global industrial agriculture system, focuses on building healthy local systems. This translates into different approaches to ecological management. For the OFA, farm ecologies are more about improved food production than land stewardship, biodiversity and soil health. The NFU on the other hand, stresses the need to balance ecological, social and economic considerations.

However, as Morgan, Marsden and Murdoch caution, "Enough has been said for us to be wary of the chain of reasoning that implies that food system localization creates socially embedded food networks that are inherently more just and sustainable than anything in the conventional system" (Morgan et al. 2006, 191). So, we also need to note that NFU farmers work off-farm to support their operations as their farms are not yet self-supporting. And, since there continues to be high demand for ecologically produced product, these farmers can get a premium for their food. However, if the few farms are to grow into a system, there needs to be an intermediary co-op or distributor to link farms together and stabilize demand. For the time being to sell their product, farmers must also have time to engage with consumers through some kind of direct sell program – either farmers' markets, CSAs or farm-gate sales. This requires both the time and ability to deal with the public, not something all farmers are suited to. Finally, there is a steep learning curve as farmers must learn to farm in new ways (Morgan and Murdoch 2001). With the small number of farmers in the network, finding mentors is a problem (Blay-Palmer 2003). Given these challenges, growing a truly alternative food *system* that is embedded in a local community is a substantial challenge.

From the Farmer Perspective

The case studies of GE/ chemical agriculture and SOD, and the OFA/NFU present aspects of the food system that have changed with time to absorb and reflect back aspects of pre-chemical agriculture (1951 census data), chemical dependent agriculture and a more recent ecologically informed, local agriculture. In the context of food systems, it is not useful to characterize shifts as opposing dualities (Sonnino and Marsden 2006) but rather as existing realities, and to understand the forces that prevent them from responding in a more ecologically sound production direction (Morgan et al. 2006; Goodman 2004). In reality then, there are OFA farmers that do incorporate some ecological farming methods into their production practices. All is not one extreme or the other. Although there are sharp contrasts between the associations, in practice there are elements of change throughout as systems include varying degrees of global/local as conditioned by nuanced social, economic and ecologically sensitive systems of food provision and production. However despite the blurred boundaries between the two groups, the reality for farmers is conditioned by their affiliations and the visions they adopt for food provisioning. It is necessary to recognize that to pit OFA against NFU farmers only helps to model salient aspects of the two approaches and as with all models, allows a clearer grasp of the most sharply contrasting points. In reality of course there are a range of hybrid systems demonstrating multiple types of embeddedness (Morgan et al. 2006; Penker 2006; Sonnino and Marsden 2006; Watts et al. 2005; Robbins 2004; Whatmore 2002). The case studies of both the SOD resistance to GMOs and the eastern Ontario farmers demonstrate that different farm groups reflect the relative importance attributed to 'local' food systems and more specifically local ecologies. The research addresses the constructed power relationships embedded in multiple scales and the way these relationships construct local food discourses and ecologies (Le Heron 2006; Morgan et al. 2006; Guthman 2004; Robbins 2004). The food re-localization project points to the capacity for more local food provisioning in the study region as well as the strong demand side interest in buying local food on the part of both consumers and retailers. So although there is strong potential demand, there is a need for increased production of market goods, as well as the creation of better links between actual production (for example, beef) with consumption. Although there is a disconnect between actual supply and demand, there is a vision from local economic development officials and the NFU to create a more localized food system.

From the vantage point of political ecology theory, the case studies point to the value of adopting a political ecology perspective. First, the research points to the contested ways in which resources are managed and the role that ecology does (or does not) have in developed world food systems. The case studies demonstrate the reach of global power structures through federal policies and the way this is translated into practice at the farm scale (McCarthy 2002; Robbins 2004). It also speaks to the value of vision as the basis for creating new spaces for local food production-consumption routines. Both SOD and the NFU provide hopeful visions and actions for sustainable farming. In keeping with Haenn (2002) the case studies expose, on the one hand how the meta-narrative is constructed and subscribed to by some farmers, while on the other hand they reveal the emerging opportunities to do food differently. In some cases farmers assimilated the message promoting global food systems and are co-creators of the decline of local ecology and local food for local eaters. In other cases, farmers are able to identify new opportunities and reframe the discourse to (re)create a vital, local food system.

The De-industrialization of Food?

We now have a sense of the historical and current regulatory, policy, marketing and retail context for the food industry in North America and the UK. In the next part of the book, we examine a watershed period in food system history – the emergence and persistence of food scares. Beginning in the 1980s, a series of food scares raised concerns about food safety and shook public confidence in the reliability of the conventional, agro-industrial food system. This shifted demand for food from industrial to more sustainably produced food. The increasing demand for trust-based, quality relationships and products are the fastest growing sector in the food industry (Minou and Willer 2003). But, we also know that agro-food companies from the industrialized food system are filling these spaces as they profit from food fears and co-opt many of the alternative food players (such as small-scale organic farmers) into the broader, agro-industrial system (Guthman 2004a, b). Recalling the example of Jolly Time candy apple popcorn, the industrialized food system plays on these consumer anxieties over health fears as a way to carve up the market and create new, niche opportunities (Schlosser 2002). The additional dimension of the highly concentrated retail market (Wrigley and Currah 2006; Wrigley et al. 2005) – for example, WalMart – will be included in our discussion of current challenges to hopeful food systems in a later chapter. In the next chapter we turn to food itself as a potent actor in the food system – that is, as a vector for disease.

Chapter 5

Translating Fear: Mad Cows, Killer Carrots and Industrial Food

The total impact of food borne disease is chilling. According to the WHO, 130 million European citizens are affected each year, while in the US 76 million fall ill annually, with 5,000 deaths. (Gold 2004)

Translation is the mechanism by which the social and natural worlds progressively take form. The result is a situation in which certain entities control others. Understanding what sociologists generally call power relationships means describing the way in which actors are defined, associated and simultaneously obliged to remain faithful to their alliances. The repertoire of translation is not only designed to give a symmetrical and tolerant description of a complex process which [sic] constantly mixes together a variety of social and natural entities. It also permits an explanation of how a few obtain the right to express and to represent the many silent actors of the social and natural worlds they have mobilized. (Callon 1986, 213)

Over the last two decades, nature has been necessarily reinserted into the industrial food landscape through a series of alarming cracks in the food system. On a regular basis there are reports of the latest food scare revealing new ways the industrial food system safety perimeters have been violated. There is a continual litany of illness, disease and death associated with eating contaminated food. In the EU, 130 million citizens are afflicted by food-borne illness annually. In the US approximately 76 million fall ill while about 5,000 die every year from food-borne sickness (World Health Organization in Gold 2004). While it is inevitable that there will be a certain amount of illness related to food consumption – and many would argue that this is a necessary part of being engaged in a natural cycle – the rates of illness found in the highly controlled food system are distressing. The question this chapter seeks to explore is, why in an increasingly regulated food system, do more and more people fall ill or die from their food every year? In trying to unravel the answer to this question, this chapter describes the emergence of industrial-based food scares since the mid-1980s including: ALAR and food born carcinogens; salmonella and eggs; BSE/ Mad Cow and Creutzfeldt Jakob Disease; and, avian flu and the threatened pandemic. The goal of this chapter is to examine some of the more well-known food scares to gain a deeper understanding about the common features that link all these food 'crises' together. Understanding the root causes of food contamination is instructive, as the links to illness and death create, to say the least, an unpredictable and ambivalent relationship between eaters and their food. As we know from the definition of fear in Chapter 1, uncertainty and a lack of control feeds anxiety.

The theoretical approach for this chapter draws upon insights from the green neo-Marxist perspective and research in Actor Network theory (ANT) (Castree 2002; Whatmore 2002). From the green political economy approach, we extract the need to make "capitalist processes and relations" "visible" (Castree 2002, 132). By tracing the power that results from capital accumulation we can understand more about how capital pushes the industrial food system in directions that run counter to social and natural well-being. As we discussed in Chapter 4 we need to understand more explicitly the forces that subordinate specific elements of society to the economy. As Carolyn Merchant (1996) explains, capitalist rationales run deep, subverting social and natural priorities to economic-capitalist ones,

> The origin story of capitalism is a movement from desert back to garden through the transformation of undeveloped nature into a state of civility and order…The good state makes capitalist production possible by imposing order on the fallen worlds of nature and human nature. Thomas Hobbes's nation-state was the end result of a social contract created for the purpose of controlling people in the violent and unruly state of nature. John Locke's political theory rested on the improvement of undeveloped nature by mixing human labour with the soil and subduing the earth through human domination. Simultaneously, Protestanism helped to speed the recovery by sanctioning increased human labour just as science and technology accelerated nature's labour. (Merchant 1996, 136)

Like Harvey (2000, 1989) and Castree (2002), we acknowledge that, "elements, things, structures and systems do not exist outside of or prior to the processes, flows and relations that create, sustain or undermine them. Here then the social and the natural, the local and the global are internally related as particular 'moments' within processes that dissolve ontological divides" (Harvey in Castree 2002, 130). Accordingly we recognize that nature and society are not separate or pre-existing. As we discussed in the context of AFS versus IFS and the idea of embeddedness, this approach creates 'false dualities' and oversimplifies the complexity of the relations (Goodman 2004; Morgan et al. 2006). Binary categories such as nature-society and global-local are oversimplified explanations of the reality of food systems (Goodman 2004, 1999). Borrowing from thinkers in urban studies, the 'actually existing' messy and constantly evolving set of networks that constitute food systems need to be recognized and unpacked (Brenner and Theodore 2002).

This is where work inspired by the Actor-Network Theory (ANT) emerges as relevant to our theoretical perspective. As Goodman (1999) explains, ANT is particularly useful in the study of food systems as it "offers conceptual and metaphorical tools that expose and address the erasures 'naturalized' by the modernist ontology" (Goodman 1999, 34). Actor Network Theory does not restrict itself to a particular lens but allows actors to describe the defining features of their relationships that emerge from their networks (Callon 1986) so that ANT "provides a way of thinking about entities and their interactions as performative, mutually constitutive and constantly emerging" (Richardson 2004, 197). The friction between actors is constantly (re)creating networks and relationships, and (re)connecting actor in dynamic networks (Whatmore and Thorne 1997). As a result, contradictions emerge in the system so that, "[f]or instance, while genetically modified foods might be the profit-led product of transnationals like Monsanto, those foods and ecosystemic outcomes can take on a very powerful, unpredictable and lively agency with real consequences" (Castree, 130).[1] By exploring and identifying the narratives and discourses constructed by the different actors in the system, we can see how these assorted players translate and manage food scares to their advantage. As Fitzsimmons (1989) challenges us,

> If what we seek is not just to understand the world, but to change it, we must address capitalism in its hidden moments, its reproduction of disguising abstractions such as "nature" and "space". Urban *(and rural)* society *(and nature)* as we know them are capitalist forms; the domination of one by the other echoes and continues the structural domination of capitalism. (Fitzsimmons 1989, 111)

To summarize then, in Chapter 5 we are interested in the way power is manifested through relational activity between different actors (including humans, technologies, institutions, and nature) and the way that these networks resolve themselves into a variety of narratives and realities about food. As two prominent US experts in food processing practices and technologies on consumer-pathogen explain,

> ...[i]n many instances it is an outbreak of foodborne [sic] illness that first alerts us to production and manufacturing processes that may be inadequate to control a foodborne [sic] pathogen. This can result in changes to the production or manufacturing process for specific products (Scott and Elliot 2006).

So, and as we shall see further, these food crises can also result in damage control by industry as narratives and realities are (re)written to serve the goals of the industrial food system. In turn these narratives can be used to recast the way we see food. This said we now turn our attention to specific instances of actors who precipitate enormous changes to the way we do food. We are especially interested in how various individual and institutional actors within the conventional, agro-food industrial system usurp nature's power by interpreting and then translating circumstances to serve their purposes (Callon 1986). In the end it seems that networks of food are mobilized to preserve networks of capital. We present the following food scares in an approximate chronological order beginning with the example of Alar as a way to document the important role that the 'intimate' and ultimately ingested commodity of food itself plays a role in shaping the industrial food system.

Managing Information: Agro-business and Translating Alar

> The Alar affair also has become a favourite media symbol for a false alarm. Reporters and pundits repeatedly refer to it as a prime example of Chicken Little environmentalism and

---

1 And while I acknowledge the importance of relationality within networks and the distributed power that results, I also concede that there is intentionality on behalf of human actors that cannot be overlooked in considering networked activity and questions of power. It is my position that intention cannot be attributed to natural actors or technologies and that this distinguishes humans from other actors.

government regulation run amok. And they are wrong. (Negin 1996, Columbia Journalism Review)

Alar is a fascinating case study of how the conventional agri-industrial sector used the media to manage risk and short-circuit consumer fears about a suspected carcinogen, Alar. Beginning in the late 1960s, Alar was used on apples to improve stem strength and colour. In the 1970s scientists found that Alar and the related chemical, unsymmetrical dimethylhydrazine (UDMH), induced high tumour formation in mice. Based on these findings, the US Environmental Protection Agency (EPA) subjected Alar and UDMH to further scrutiny. It is at this point that the dispute begins. In 1984 the results from the review prompted EPA scientists to recommend removing Alar as a chemical for use in food production. The Scientific Advisory Council of the EPA, however, recommended allowing a lower level of Alar to be used. In the end, the Council decision, and not that of the scientists, was upheld. The decision to side with the Advisory Council and not scientists was to repeat itself in future food scares including threats posed by salmonella in eggs and Mad Cow disease in the beef system.

By 1989, the EPA had evidence that cancer risks for adult exposure to Alar and UDMH was 5 cases per 100,000, well in excess of the EPA acceptable rate of 1 case per million. Importantly, this evidence was for adults and ignored children. This is crucial as children are the major consumers of apples and apple juice and so are exposed to the chemicals at much higher levels. Children's bodies are also still forming and growing and so are generally even more susceptible to the effects of chemicals than adults.

While the EPA was assessing the situation, an environmental group called the National Resources Council (NRC) launched its own research. This resulted in the report 'Intolerable Risks: Pesticides in Our Children's Food'. The report revealed that,

> ...[t]he average preschooler's consumption of food contaminated with UDMH would cause one extra cancer case for every 4,200 children exposed during their first six years of life-a level of risk 240 times higher than that considered acceptable by EPA following a full lifetime of exposure. (Rodgers 1996: 178)

Given the importance of the report for children's health, the NRC hired a public relations firm to manage the release of their findings. First the report was made public on a CBS *60 Minutes* program that was viewed by over 40 million people. Following the *60 Minutes* show, Meryl Streep decided to become a spokesperson and held press conferences, TV interviews including spots on *Donahue* and the *Today Show*, as well as appearing in a public service advertisement. The public response was rapid and apple and apple juice sales plummeted; the apple industry was perched on the edge of disaster (Rodgers 1996). By the end of 1989 the EPA banned the use of Alar for food production. Although by the fall of 1990 the apple industry rebounded (Negin 1996), Alar had become symbolic for the agri-food industry on how NOT to manage a food crisis. Notwithstanding the fact that the Alar/apples battle had been 'lost', the larger food industry – interested in the continued use of chemicals – would learn and react.

From a communication studies perspective, the food industry's management of the issue of Alar and its association with apples in the wake of the 'Intolerable Risks' report is instructional as it illustrates the way risk is constructed and translated by different actors along the food production-consumption chain. Following the CBS program, the apple industry went on the offensive and successfully reversed the media stance and then public opinion about Alar in a bid to undermine negative perceptions specific to the use of chemicals in food production and more generally to the use of science and technology in food production. Although the apple industry never regained the right to use Alar, the food industry more broadly was adamant in their conviction that they would drive public opinion and hence policy in *their* area,

> ...[a]ccording to John Stauber, editor of the Madison, Wisconsin-based newsletter *PR Watch*, the erroneous reporting on Alar [that Alar is harmless] is largely due to a sophisticated public relations counterattack mounted shortly after the *60 Minutes* show. The controversy "scared the hell out of the agribusiness and food industries", he says. "The food industry said, 'Never again', [and] set out to convince the news media this was a hoax". The campaign, he adds, has been "very successful". (Negin 1996)

In lockstep with other corporate interests such as tobacco companies and the biotechnology industry as discussed in Chapter 2, the American Council on Science and Health (ACSH) had been established in 1978 to lobby the government and provide 'educational' material to the public.[2] ACSH was funded by companies such as Uniroyal – the manufacturer of Alar – who contributed $25,000 to its annual budget. The goal of the ACSH was to promote a perspective that discredited naysayers of chemical use in agriculture. For example the ACSH contended,

> ...that animal tests cannot prove a product's carcinogenicity, and that NRDC was crying wolf. Once these initial stories entered into electronic databases such as Nexis, Datatimes and Dow Jones that journalists routinely use for research, they were repeated uncritically and uncontextually in subsequent stories. This further stacked the database deck for reporters too lazy to find the most credible sources. (Fulwood 1996, 12)

As the popular media rely increasingly on experts including those in advocacy roles such as the ACSH for information, industry is able to translate risk into relatively more favourable terms and deflect consumer attention away from potentially pressing food safety issues.

A 1993 report by the National Academy of Science concluded that existing federal laws do not adequately protect infants and young children from dietary exposure to chemicals. The chair of the committee that issued this report, Dr. Philip Landrigan said that, "NRDC was absolutely on the right track when they excoriated

---

2 Other American organizations like the Alliance for Environment and Resources and the Abundant Wildlife Society actually support clearcut logging, and trapping and shooting of coyotes, mountain lions and other predators, respectively. Responsible Industry for a Sound Environment is a pro-pesticide group founded by the American Crop Protection Association. A Canadian example of this is the 'wise use' movement among Canadian timber companies.

the regulatory agencies for having allowed a toxic material such as Alar to stay on the market for 25 years without proper toxicity testing" (Fulwood 1996,11–12).

The UK and Salmonella: Changing Food Politics

The story of salmonella in the UK is instructive as it exposes the way that weaknesses in regulation and policy can be self-perpetuating when there is a lack of checks and balances in government. The salmonella-in-eggs crisis in the late 1980s exposed a regulatory system that listened exclusively to industry interests. It is also useful as it points out the possibility for institutional change when a broader scope is adopted.

Salmonella has been a public health issue in the UK and elsewhere throughout the twentieth century (Hardy 2004). Between 1995 and 2004, nearly 500,000 cases of salmonella were reported. The illness killed 119 people in 2000 alone. Despite the established and on-going risk linked to salmonella, it was the salmonella-in-egg scare in the late 1980s that marked a policy watershed in the UK as food politics began to take consumer input into account. Before the salmonella-in-eggs incident, decisions related to food safety were made by 'experts' including farmers behind closed doors in the name of the public good. In fact, from the 1950s up to the late 1980s, UK farmers and food regulators decided unilaterally about food availability, production, and safety standards. This decision making process arose from the priorities and circumstances that faced the UK during WWII when the Ministry of Food was created. As explained in Chapter 2, food quantity was targeted as a key priority, so that,

> The government recognized that because it has a role in *ensuring* a nutritious diet, the structure of agriculture should be economic... (Smith 1991, 236–237 emphasis in original)

In other words farmers were instructed to produce large quantities of food and they were assured that they would be compensated for their efforts. The post-WWII world food shortage and the UK financial crisis that led the government to make domestic food production a priority, helped to forge strong ties between the National Farmer's Union (NFU) and Ministry of Agriculture, Fisheries and Food (MAFF). This laid the foundation for a network of experts that excluded consumers; the assumption was that consumers wanted more food, and that their interests were in line with those of farmers. Decision-making proceeded accordingly.

Under this regime of accelerated food production, policy makers were entrusted with a technical duty to ensure food safety. Consumers were advised how to safely prepare their food and were accountable for proper food handling but were not consulted about the determination of food safety requirements as they affected production and distribution. As discussed in Chapter 2, silos were created that carved food up into safety, health, production and nutrition issues. Parliament was also excluded from the specifics of the process. Having established the need for food safety standards, the government left food safety details to the bureaucrats. As a result, three expert communities emerged, centering on the following food chain dimensions: 1) food production; 2) health and diet including medical issues; and, 3) food safety.

Eggs as a vector for salmonella in the late 1980s shifted this policy-making approach as the regulatory bodies were caught off guard by an inexplicable rise in salmonella related illnesses in the human population. Uncertainty and buck passing between the ministries of agriculture and health, and the refusal by the egg industry to admit a link between salmonella and eggs meant the crisis deepened as it was 'managed'. The result was a shift in power so that,

> The *salmonella* crisis was indicative of a general weakening of the position of the farmers partly as a result of long-term changes in the food policy community. The farmers failed to keep the *salmonella* in eggs issue off the agenda. The policy community which had previously managed to avoid conflict was now subject to widespread political debate. Food poisoning was transformed from an issue of a technical nature and of individual hygiene to one of central political importance. (Smith 1991, 244)

Eventually the link was made to eggs as the source of *salmonella*. However, while regulators struggled to establish the cause, the genie escaped from the bottle (Smith 1991). Faced with a near tripling of UK *salmonella* cases, the Health Minister Edwina Currie stated, "Most of the egg production in this country sadly is now infected with *salmonella*" (Hickman 2006: 1). Pandemonium ensued and egg sales dropped 60% by the next day. To stem the flow, four million hens were slaughtered, 400 million eggs were destroyed, the government compensated egg farmers for their losses, and Ms Currie resigned her position as Minister of Health (Hickman 2006).

Apart from the economic and political upheaval, of particular interest for our purposes is the food scare as an opportunity for consumer concerns to be heard and inserted into the political process. Salmonella in eggs opened a space in the existing community of farmers, scientific and policy experts to give voice to the previously unheard consumer (Smith 1991). Here we have what Lee and Leyshon (2003) refer to as the 'crack in the neo-liberal sidewalk' as the issue of salmonella in eggs pried open a space for consumer activism in the UK. While the government lagged in developing meaningful standards, previous relationships broke down and new ones emerged. One consequence was that actors further along the food chain undermined the power of farmers.

The opening to consumers came from food retailers as several circumstances dovetailed with the eggs-in-salmonella crisis to change food industry dynamics. First, food manufacturers had been poised to bargain farmer profits down when UK price protections were lifted as a condition for the UK to join the EC. Second, for over two decades, manufacturers and retailers had been consolidating operations as a way to increase profits. Between 1961 and 1982 the number of small retailers decreased by 50% while the number of supermarkets increased by a factor of eight (Gardener and Sheppard 1989 in Smith 1991). And third, as part of their bid to grow their businesses in this relatively stable retail environment, retailers assumed the mantle of 'defender of the consumer' by introducing safety and handling standards for production and processing that exceeded those established by government bodies. They also developed and adopted labelling and additive free standards that they imposed on food processors. As Smith remarks,

The retailers have to an extent undermined the policy community partly through representing interests in conflict with those of farmers and the food manufacturers but more importantly through by-passing the community completely. Hence, they have reacted to the market power of the consumers because of the competitiveness of retailing and in a sense introduced a new form of political action into the food arena. Food retailers are strong enough to develop 'private policy' in concert with their consumers and thus short-circuit the constraints and interests of the policy community. Thus they can impose their own standards on producers and manufacturers regardless of government policy. This has resulted in them taking a lead in areas of food policy. Tesco's healthy food campaign was a means of encouraging healthy eating and not just a reaction. The retailers' demand for better hygiene has the impact of raising the issue of food handling and so forces the government into taking the issue seriously. (Smith 1991, 248)

One could argue that this established the foundation in the UK for recent activism as consumers tore out GE crops, demanded more accountability and transparency in the food system, and most recently as they demand to know more about the ecological impact of the food they eat (BBC 2007a).

However these are certainly only incremental differences and, it is important to continue to be critical in assessing the changes that ensued from the salmonella incident and to recognize the weaknesses that remained in place, and the negative precedents that were set. First, in a defensive move from the factory farm chicken industry, part of the blame for the crisis was attributed to free range egg producers. This foreshadows reactions from industrial actors in other food scares when the industrial food system deflects blame onto ecological producers. We will deal more with this in the section that describes the avian flu pandemic. Second, the lessons from food scares are not easily learned and are unfortunately repeated. As O'Brien reminds us,

In 1979, ahead of the BSE crisis, the Royal Commission on Environmental Pollution warned against the dangers of feeding dead farm animals to live ones, highlighting the possibility of recycling disease-causing agents. And yet, even now, nearly 20 years after this warning from the Royal Commission and 10 years after the farming industry became aware that the BSE crisis was triggered by feeding dead animals back to live ones, we continue to make cannibals of farm animals: poultry can still be fed with hydrolysed feather meal, and the 'off-cuts' and waste blood from poultry abattoirs. (1997, 7)

Although some change was realized, many questionable practices emerged and some remained following the prevalence of salmonella-in-eggs. For example, high-density chicken rearing methods continue to create conditions for the spread of salmonella when poultry feces come into contact with other poultry due to crowded living and transportation conditions. The breeding of chickens to produce the maximum amount of breast meat means that as chickens fatten they have difficulty supporting their own weight and often rest their breasts on contaminated floors or cages surfaces (O'Brien 1997, 14). Although salmonella was brought under control by a program of vaccines in the mid-1990s, this is a temporary fix as it exposes birds and humans to other threats as bacteria mutate and develop immunities to antibiotics. As we see in the next section, the case of salmonella foreshadowed many features of the UK BSE crisis that emerged in the 1980s.

Mad Cows

Mad Cow disease or Bovine Spongiform Encephalopathy (BSE) was first recognized in UK cattle in 1986. As a brain wasting disease, it is believed with almost 100% certainty that BSE is caused by prions that reside in the brain and spinal tissue of infected animals (Center for Disease Control and Prevention 2007). It is thought that the defective prions highjack normal protein formation in the brain and cause a degenerative condition that ultimately results in death. Sarah Whatmore (2002) describes the power of 'rogue' prions and the effect the Mad Cow crisis wrought on the industrial food system in the UK and beyond. According to Whatmore, prions connected "the sites and practices of food production and consumption, animal and human well-being". Prions were,

> ...unlikely allies in undermining the prevailing commercial, policy and scientific cartographies of affectivity and responsibility and making space for more relational ethical possibilities. (Whatmore 2002, 164)

Prions wreaked havoc on the international beef food chain and forced a reconfiguration of the actors and networks in the industrial beef system.

The force of prions was only gradually understood, and then only reluctantly admitted in the UK. Although the first cases of BSE were detected in cattle in 1986, it was not until 1996 that the connection between BSE and the human variant Creutzfeldt-Jakob Disease (vCJD) was established in the UK. At that point over 700,000 infected animals had entered the UK human food chain. As a result, over 4.5 million animals were destroyed as UK officials attempted to reassure domestic and international beef eaters about the safety of UK beef (Weiss et al. 2006). By 2005, 152 human deaths had been attributed to vCJD (Gale 2006).

The UK Government was criticized heavily for the way it communicated human health risks arising from BSE exposure. In a situation that recalls the UK government response to salmonella-in-eggs, the government was reluctant to admit links between BSE and vCJD based on advice from a working group composed of members from the 'Southwood Working Party'. This group was convened in May 1988 and included: experts in medicine, in particular in neuropathology and Spongiform Encephalopathy; veterinary scientists; medical and scientific officers; officials from the MAFF; licensing officials; and the Minister of Health. This group determined the risk of human infection to be 'remote' (Klint-Jensen 2004). In November 1989, the

> MAFF wanted to avoid "scare stories" (BSE Inquiry 6: 4.483) and made arrangements for the reporting of these studies in such a way that a press release could say "that the results of these experiments add to the evidence that we are dealing with a disease similar to scrapie,[3] and restate the fact that there was no evidence that scrapie is transmissible to humans". (BSE Inquiry 6: 4.485) (Klint-Jensen 2004, 414)

---

3 Scrapie is the disease in sheep that is comparable to BSE crisis in cows. Some suggest that contaminated brain matter from scrapie infected sheep contaminated cattle feed in the UK and led to the BSE (Rothstein 2004).

The communication of risk associated with human transmission erred on the side of reassuring the public, not protecting it. The goal was to preserve the industry not the consumer. The important effect of this was to dampen the sense of urgency on the part of the people charged with enforcing orders to remove offal from the food chain as a means to reduce the risk of infection (Klint-Jensen 2004). Priority was also accorded to the protection of export markets and the farm industry (Phillips et al. 2000),

> Initially MAFF only banned the use of offal in baby foods and originally refused to make BSE a notifiable disease on the grounds that there was no proof that the disease could spread to humans. It was only eighteen months after the disease was discovered that [the] Minister of Agriculture ordered that all cattle with BSE be destroyed (*The Guardian*, 11 July 1988) and only in June 1989 that the use of offal was banned (*The Guardian*, 12 June 1989). This apparently demonstrates MAFF's continued willingness to side with the producers when there is doubt". (Smith 1991, 252)

"The lowest point in the history of food regulation can be dated with some precision: it was March 20, 1996" (Morgan et al. 2006, 47) when the Government admitted a probable link between BSE and vCJD. After an entire decade had passed the connection between BSE and vCJD was finally acknowledged. With this revelation, public trust in the political system was violated and BSE became a,

> ...vivid example of commercial greed, of inept food safety authorities, untrustworthy politicians and a nature that, according to media coverage, had the ability to 'strike back'. (Lien 2004, 3)

And so, in keeping with circumstances surrounding other food scares, the prion was one catalyst that pushed the reassessment and revamping of the UK regulatory and food safety system making it more transparent and reflexive. A cornerstone of the restructuring was the disbanding of the MAFF and the creation of the Department of Environment, Food and Rural Affairs (DEFRA) and the Food Standards Agency (FSA). "Charged with the twin tasks of rebuilding public trust in the food supply chain as well as in the food governance system, the FSA sought to create a new consumer-friendly regulatory regime based on three core goals: to put consumers first; to be open and accessible; to be an independent voice" (Morgan et al. 2006, 49). The FSA is 'innovative' as it: 1) reports directly to Parliament and the devolved assemblies; 2) is composed of a wide cross section of experts from public health, consumer groups, food industry including processing and services who were 'appointed after a free and open public competition'; and 3) all meetings are held in public and any member of the public can put questions to the Agency (Morgan et al. 2006: 49). Other innovations include the examination of several interconnected issues on food system sustainability as well as a conscious attempt to break down the silos between departments. These initiatives added vigor and credibility to the new UK food regulatory regime. It seems that perhaps the UK regulators are now in tune with public food safety goals (Marsden 2007).

    If we now consider the North American version of the BSE crisis, the approach to disease management is worrisome particularly when viewed in light of the UK experience. It would appear that the 'lesson learned' in North America has been to try to lower international standards as a way to make North American beef exportable to international markets and not to improve the food safety system for consumers (Weiss et al. 2006).

The North American BSE story begins in 2003 when the first case of Canadian BSE was detected in a cow in Alberta. The source of BSE was traced to animal feed. When the announcement of BSE was made, the US, Mexico and Japan closed their borders to Canadian beef.[4] This was especially damaging to Canadian farmers who had taken the advice of their federal and provincial governments and oriented themselves to serve the US market. As noted in Chapter 4, this move occurred after the opening up of the Canada-US border following the implementation of the North America Free Trade Act (NAFTA) (Blay-Palmer and Donald 2007). As a result, Canada had structured its beef industry with diminished processing capacity in Canada so that live cattle were moved into the US and processed there. By 2002 the export of beef and live cattle was valued at $4 billion. Seventy percent of this trade was with the US (Oliver and Fairbairn 2004). The closure of the US border to live animals and meat from Canada was a devastating blow to the industry. At the peak of the crisis, Canadian cattle producers were losing up to $11 million per day and suffered a loss of 5,000 jobs.[5]

The next blow came on December 23, 2003 when the US announced its first case of BSE. Within days, the cow's birthplace was traced to Alberta, shifting the blame from the US onto Canada. However, the crisis continued to spread so that by August 2006, the US had reported 2 cases of BSE while Canada had reported its eighth case. Given the widely understood and broadcast UK situation, a question of central interest for this book is how did these animals become infected with the BSE prion when so much was known about the UK case? To answer this question, we need to understand the actions of Canadian and US regulators as they managed and interpreted the BSE situation over the previous two decades.

Several regulatory and institutional changes over the course of several decades converged to precipitate the BSE situation in Canada and the US. The changes made were consistent with an increasingly industrialized global food system. They created a progressively complex food system that in the end was so divorced from nature we ended up turning ruminants into cannibals. A good place of entry into this tragedy is the restructuring of Canadian institutions as part of NAFTA.

In keeping with the move to a free trade environment, provincial and federal governments not only encouraged food producers to fully integrate themselves into the US market but also restructured their own departments and ministries with a view to increasing food exports. Federally, as departmental responsibilities were changed and new institutions created, the Canadian Food Inspection Agency (CFIA)

---

4 Japan reported 20 cases of BSE between 2001 and mid-2005 (CBC 2006b).

5 Anecdotally, increased farmers suicides related to BSE are reported for Canada (Kilgour 2004) during this time. In Manitoba, calls to the Farm and Rural Stress Line increased from 863 in 2002 to over 1400 in 2003. Calls directly related to BSE for 2003 and 2004 totalled over 280 (Manitoba Farm and Rural Stress Line 2005). In the US farmer suicides are three times the rates in the general population.

was tasked with the monitoring food safety in Canada. But, leading up to and during the BSE crisis, the Canadian Food Inspection Agency (CFIA) was also tasked with promoting export opportunities for Canadian agriculture making it both the industry watchdog and cheerleader. Provincial ministries were also restructured as part of the move to a free trade environment. As mentioned in Chapter 4, in the mid-1990s the Ontario Ministry of Agriculture, Food and Rural Affairs (OMAFRA) diminished its role as a strong supporter of farmers in providing services directly to farmers, and moved instead to develop commodity-based export markets. In these contexts, profit assumed a very prominent role in decision-making (Blay-Palmer 2003).

Within this changed regulatory context, in 1997, recognizing the potentially devastating effects of a BSE incident, the Canadian government imposed a ban on rendered beef products in cattle feed. However, this ban was not extended to feed for poultry and swine until later. Inevitably, BSE infected feed actually destined for poultry or swine made its way into the cattle feed stream and infected the cow in Alberta. Despite the knowledge about the economic and human health risks associated with BSE from the UK experience, the beef industry and Canadian regulators have not acted in the best long term interests of the industry or Canadians. In fact it is only recently that the ban on all rendered animal parts in all animal feed was brought into force effective July 12, 2007 (CFIA 2007a). As well, blood products – known to transmit BSE (United States Department of Health and Human Services 2004) – are exempt from the ban (CFIA 2007a). Although figures for blood products used in animal feed are not available for Canada, the US rendering industry reports a total of 226.5 million pounds per year of blood products that are permitted to be used in the animal feed system – of this 121.9 million pounds come from ruminant animals[6] (Hamilton et al. 2006, 84).

The Canadian federal government recently recognized the conflicting CFIA mandates of economic promotion and food safety. As of 2007 the CFIA was revamped so that it is now responsible for food aspects of Acts related to food and drugs, agricultural products, meat and fish inspection, consumer packaging and labeling, plant protection, animal health, food inspection, seeds, feeds and fertilizer (CBC 2007). The CFIA mandate is to:

1. Protect Canadians from preventable health risks
2. Protect consumers through a fair and effective food, animal and plant regulatory regime that supports competitive domestic and international markets
3. Sustain the plant and animal resource base
4. Contribute to the security of Canada's food supply and agricultural base, and
5. Provide sound agency management (CFIA 2007b)

But, the economic development goals implicit in the second mandate item raise concern that conflicts of interest have not been entirely eliminated. As well, since the CFIA president reports to the federal minister of the Department of Agriculture and Agri-Food Canada (AAFC) whose mandate is focused on export development

---

6 "Rendering adds nearly $1 billion in value to the U.S. livestock production sector in the form of proteinaceous feed ingredients alone. This value approaches $2 billion when contributions from rendered fats and greases are also considered". (Hamilton et al. 2006, 73)

(recall the economic development focus in the government consultation documents discussed in Chapter 4) there continues to be the potential for conflicts between economic development and the imperative for safe food. Finally, the presence of animal blood in the food chain despite connections to BSE transmission raises questions about CFIA's ability to make independent choices about food safety for Canadians (CFIA 2007a).

In addition to provoking questions about food safety, the Canadian BSE crisis also raises concerns about the nature of competition in the beef industry. In the words of the House of Commons Standing Committee on Agriculture and Agri-Food,

> In the aftermath of the BSE crisis, many industry experts and the public at large have observed cattle prices plummeting well below economically viable levels for many cattlemen. At the same time, the wholesale and retail prices of beef products have either risen or fallen by a much smaller proportion than cattle prices. The growing spread between farm-gate and retail prices has led many industry observers to express concern over the recent consolidation and rationalization within the packer and processing segment of the industry, which may have resulted in too much concentration of ownership. (Steckle 2004, 3)

Prices paid to farmers for their steers averaged $1,448 between January and May 20, 2003. Between May 20 and the end of December 2003, steer prices paid to farmers averaged $857, a drop of 40% from the pre-BSE prices. Retail prices declined only 14% between May and September 2003 and then trended up. The share of steer prices accounted for 15% of the retail price of beef and only 7% at the peak of the crisis down from 25% in 1999. A price squeeze has been increasing for farmers as consolidation has accelerated in the packing industry. For example, in Alberta the heart of the Canadian beef industry, between 1991 and 2003 while the number of feedlots fell from 229 to 212, the number of cattle processed increased from 927,000 cattle to over 2 million. The average feedlot has grown from just over 4,000 animals to over 11,500. Feedlots with over 10,000 head accounted for 59% of production by 2003, up from the 31% levels in 1991. Increased consolidation is also occurring in the packing industry. The western segment of the industry is dominated by three companies – Cargill, Lakeside (a division of Tyson) and XL Beef – who process 95% of western beef (Steckle 2004).

While Canada developed a program to cope with BSE, the US adopted its own response after BSE was first detected there in December 2003. First, the US increased the number of animals tested for the disease. Over 759,000 animals were tested over a 25 month period beginning in 2004. The accelerated testing was implemented "with the aim of determining the disease's prevalence" (CIDRAP 2006). Since the end of the accelerated testing program, the USDA has returned to pre-BSE detection standards of testing 0.1% of all animals slaughtered. Second, the US lobbied to change international standards to keep them in line with the reality of the US beef industry (Weiss et al. 2006). The following description of the current mandate for the OIE (Office International des epizooties - World Organization for Animal Health) outlines the rationale for an international, intergovernmental body to recommend global standards for animal health and food safety (OIE 2006),

As a result of globalisation and climate change we are currently facing an unprecedented worldwide impact of emerging and re-emerging animal diseases and zoonoses (animal diseases transmissible to humans). Improving the governance of animal health systems in both the public and private sector is the most effective response to this alarming situation. (Vallat 2007)

Compliance with OIE standards is voluntary. There are also questions about credibility. For example, following the detection of BSE in the States the US lobbied the OIE to make important changes to BSE standards to be in line with US compliance capability. By 2005 the standards changed so the category 'BSE free' was subsumed under the less stringent classification of 'Negligible BSE risk'. The elimination of the BSE-free category privileges countries like the US and Canada where BSE incidences have occurred. It also makes the system more difficult to follow in all its details. According to OIE guidelines the 'Negligible BSE risk' category indicates cases where there are adequate systems to monitor the flow of hazardous spinal cord and brain material to ensure there is no risk of contamination. So the US is allowed to claim the 'Negligible BSE-risk' status, even though it continues to allow the feeding of animal parts to poultry and swine (Weiss et al. 2006). This offers the opportunity for feed with SRMs to get into the cattle feed stream as we have seen in Canada.

The US approach to BSE may compromise food safety on a global scale. Australia is an excellent case in point. As of 2006, Australia was BSE-free thanks to feed standards implemented in the 1960s that precluded beef consuming protein from other ruminants. However, under pressure from US controlled Australian beef processing companies and supported by an Australian-US Free Trade side-deal, the Australian beef market and its consumers may now be exposed to BSE as Australian borders open to international beef products (Weiss et al. 2006). In 2006 the OIE reported that Australia was BSE free (Government of Australia 2006; OIE 2006).

The Avian Flu Pandemic

The threatened pandemic reveals fascinating tensions between the global poultry trade and small-scale family farmers around the world. The struggle, that calls to mind 'David and Goliath' parallels, helps to crystallize several key points of conflict between the two types of production.

Initially, the spread of avian flu was blamed on wild bird migration and backyard poultry flocks. As a result of this interpretation, industry and regulators destroyed populations of poultry en masse, large and small flocks alike in order to halt the spread of the disease. Over 17 million birds were destroyed in the US between 1983 and 1984 in an avian flu incident. In British Columbia, a 2004 outbreak resulted in the destruction of nearly 15 million birds including 553 small flocks with rare genetic diversity (a total of 18,000 birds were culled) and 410 commercial poultry farms (14.9 million birds) (Hudson and Elwell 2004). In reporting on the BC response to the 2004 outbreak, a CFIA veterinary epidemiologist commented that to prepare for the possibility of other outbreaks, "the poultry industry must adopt a fortress mentality" (Hudson and Elwell 2004, 13).

The strong defensive attitude stems in part from the threat that a pandemic poses to human health. The most recent and widespread example of a global flu is the 1918/1919 Spanish flu pandemic that killed an estimated 50-100 million and left no continent untouched. While some countries race to stockpile vaccines to avoid a similar global health catastrophe, organizations such as Canada's Poultry Industry Council urge farmers to increase their bio-security measures and keep poultry flocks away from any potential contamination sources. But this response appears to treat only the superficial vectors for the spread of a potential pandemic and ignores the underlying causes. In fact, it seems that the poultry industry is responsible for creating the ideal circumstances for the propagation and spread of a pandemic. There is a growing consensus that the homogeneous genetic material, movement of chickens and their by-products within the food system, crowded and often poorly ventilated living conditions, and the global reach of the poultry industry are the factors that could push us into a human health crisis.

First, broiler[7] poultry flock genetic material is almost identical – so, for example, the sperm from one rooster can be responsible for over 2 million offspring. In a 2004 study of broiler living conditions in the UK, 75% of flocks were of the same breed (Stamp Dawkins et al. 2004). This extensive genetic consistency is needed to feed the automated food processing and food services industry where chicken parts must be as similar as possible. As indicated earlier, chickens are bred for maximum breast meat in the shortest time. To achieve these efficiencies, there is little room for genetic differences. If you are a chicken processor, this makes sense. If you are a flu virus, this is the ideal territory to infect and spread throughout a poultry barn as once you have infected one bird, the rest are all the same.

In addition to the lack of genetic diversity is the robust nature of viruses that allows them to spread from location to location as animals and their waste move through the food chain. For example, the virulence of flu viruses means that they can survive in fecal matter or water used in aquaculture where poultry protein may be used as a supplement. The virus can live for up to 35 days in these conditions (Lucas and Hines 2006). The connection between poultry barns and aquaculture provides a link that could spread the virus beyond one large-scale operation. The next step is the jump into a host animal species that shares viruses with humans. Pigs are the perfect vector (World Health Organization 2007). The practice of raising pigs and poultry on proximate industrial scale farms create the ideal jumping off point from poultry to swine to humans (Lucas and Hines 2006). Although the industry is quick to reassure the international community about the safety of their handling procedures, avian influenza outbreaks continue with over 15 since 1976 in many countries. Australia reported outbreaks in 1976, 1985, 1992, 1994, 1997; the United States in 1983, 2002, 2004; Great Britain in 1991; Mexico from 1993 to 1995; Hong Kong in 1997; Italy in 1999; Chile in 2002; Netherlands in 2003; Canada in 2004; Azerbaijan, Turkey and Iraq in 2006; Cambodia and Indonesia in 2005, 2006 and 2007; China in 2003, 2005, 2006, and 2007; in Djibouti in 2006; Egypt in 2006 and 2007; Nigeria and the Lao People's Democratic Republic in 2007; Thailand annually

---

7 Broilers are chickens raised for meat, while battery refers to the way that egg-laying hens are raised.

from 2004 to 2006; and, Viet Nam annually since 2003 except for 2006 (WHO 2007, Lucas and Hines 2006). Since 2003 there have been 329 human cases of H5N1 with 201 deaths (WHO 2007).

International guidelines established by bodies such as the World Health Organization and the UN (WHO 2007, Webster and Hulse 2004) direct farmers to keep poultry indoors, while the Canadian Poultry Industry Council,

> ...is urging farmers to be even more conscientious than usual to protect their flocks...and ultimately, to protect the humans who come in contact with them. Producers normally adhere to strict measures to maintain clean barns, and keep out avoidable germs, bacteria and viruses. This is why casual walks and tours by humans through chicken barns are discouraged – random access creates too much potential for introducing disease. (Canadian Poultry Industry Council 2005)

However, keeping birds inside may in fact add to the problem. Reports from the Humane Society of the United States and GRAIN[8] indicate that confined living conditions typically found on factory farms promotes disease in broiler chickens (Humane Society of the United States 2006, GRAIN 2006). In a study of poultry density and living conditions in intensive broiler operations in the UK, Stamp Dawkins et al. (2004) reported that in their study group the size of chicken houses varied between $455 - 1,900$ m^2 with up to 53,000 hens per house. The stocking densities that were studied ranged from $30 - 46$ kg/ m^2 with the average bird weighing between 2 and 3 kilograms when slaughtered. Clearly, chickens living on factory farms live in cramped quarters.

Additionally factory farms are not impenetrable despite bio-security measures that include the use of masks, chemical baths, and showers for employees to maintain sterile environments for the poultry. Studies from the FAO, the EU and Japan point to high levels of dust and insects associated with factory chicken farms. Both can act as vectors for the spread of disease (Nierenberg 2007). The shipment of live birds, poultry products and contaminated clothing from factory workers are other ways of spreading H5N1,

> ...the huge concentration of CAFOs [Concentrated Animal Feeding Operations] and CAFO workers in certain areas creates further opportunities for the disease [avian flu] to spread. According to the FAO report, when CAFO workers make up more than 15 % of a community, they may act as avian flu amplifiers for the community as a whole. Small and medium-sized farms, on the other hand, can often prevent diseases among poultry and other animals more efficiently. Such farms usually have more genetically diverse flocks and herds that are less stressed and better adapted to climate and disease. (Nierenberg 2007, 1)

Global trade may also contribute to the spread of an influenza pandemic. The presumed source for the threatening pandemic is the H5N1 variant of avian influenza that first appeared in 1996 in southeast China. Since then it has emerged in other parts

---

8 "GRAIN is an international non-governmental organisation which promotes the sustainable management and use of agricultural biodiversity based on people's control over genetic resources and local knowledge" (GRAIN 2007).

of Asia, Africa and the EU. Contrary to widespread impressions, avian flu does not follow migratory bird patterns. French ecologists Gauthier-Clerc, Lebarbenchon and Thomas, "conclude that human commercial activities, particularly those associated with poultry, are the major factors that have determined its global dispersal" (British Ornithologists Union 2007, 1). Recent movement of the disease helps to clarify how experts think the virus spreads. Following the discovery of hundreds of dead migratory wild birds in 2005 on Lake Quinghaihu in China, the infection was traced to several domesticated sources including: 1) "local poultry farming (on the lake shore and surrounding area) or, 2) from the release of captive-bred Bar-headed Geese reared to supplement the wild population in order to maintain a high level of higher quality poultry meat for harvesting and human consumption" as well as to multiple local chicken factory farms; 3) the use of chicken feces in nearby fish ponds; and 4) train and road links to a Tibetan site of bird flu (GRAIN 2006; BOU 2006). In fact, it now seems that commercial operations are threatening wild bird populations as industrial barns provide the conditions for the virus to become more virulent and change from the less threatening low pathogenic avian influenza (LPAI) into high pathogenic avian influenza (HPAI) (Lucas and Hines 2006). According to a report authored by an internationally-respected NGO,

> Within crowded poultry operations, the mild virus evolves rapidly towards more pathogenic and highly transmissible forms, capable of jumping species and spreading back into wild birds, which are defenseless against the new strain. In this sense, H5N1 is a poultry virus killing wild birds, not the other way around. The same argument holds for small-scale poultry production. Bird flu does not evolve to highly pathogenic forms in backyard poultry operations, where low-density and genetic diversity keep the viral load to low levels. Backyard poultry are the victims of bird flu strains brought in from elsewhere. When backyard farms are separated from the source of highly pathogenic bird flu virus seems to die out or evolve towards a less pathogenic form. (GRAIN 2006, 8)

The global food industry and the movement of food products is also a pathway for the spread of disease. For example, a recent outbreak in Suffolk UK confirmed the transmission of H5N1 through global trade in poultry rather than through wild birds (Gauthier-Clerc et al. 2007, BOU 2007). This case is worthy of more attention as it documents the details of the case as a matter of public record through the DEFRA epidemiological report submitted in April 2007. The site in question is a vertically integrated facility that includes turkey processing, slaughter and abattoir facilities. Poultry feed is supplied to the broiler barns from an off-site facility owned by the same company. At the time of the infection, the turkey processing part of the operation had 159,000 turkeys on site housed in 22 barns (an average of over 7225 birds per barn). Standard operating procedure for this type of bio-secure facility was adhered to including the use by each employee of two pairs of boots (one to enter the barn area and another to enter the barn itself), multiple chemical bath points, the use of face masks and the separation of staff working in different site divisions (i.e. employees work in either processing, slaughtering or abattoir). The final conclusions of the report stated, contrary to theories that the virus came from wild birds in continental Europe,

There was no evidence to support the hypothesis that wild birds were the source of the outbreak. This was based on the fact that there had been no isolations of H5N1 from wild birds in Europe during the 2006/7 wild bird migration period and subsequent residency. There had been no weather induced movements of wild birds from mainland Europe as occurred in the early part of 2006 which could provide a link between Hungary and Great Britain, especially as the first outbreak in Hungary and that in Suffolk occurred within a short time of each other and no common source of migratory birds was evident. (emphasis added, DEFRA 2007, 21)

DEFRA concluded that the source of the infection was from a Hungarian turkey processing plant that shipped meat to the UK turkey processing plant in January 2007, "Our conclusion is that infection was most likely introduced to GB via the importation of turkey meat from Hungary" (DEFRA 2007, 30).

As well, research by experts associated with the British Ornithological Union reported that the H5N1 pathways do not correspond to wild bird migratory routes. Even when migratory birds are infected only small numbers die (BOU 2006). Finally, H5N1 has been transmitted to humans exclusively from factory farming sources. According to the BOU (2006),

The spread of H5N1 from Asia westwards into Europe does not follow natural migration routes of wild birds. The spread follows an east-west direction cutting across the north-south flyways of wild birds. The main line of spread of H5N1 closely correlates with the main east-west transportation routes, human movement and habitation. In Africa, the isolated and highly separate cases are indicative of Man-moved poultry and not movements of wild birds. Nigerian officials were quick to announce that their confirmed cases were caused by the illegal importation (from the Far East) of poultry. (BOU 2006)

Commenting specifically on the role that the movement of wild birds plays, experts from the BOU assert that the wild bird population that does get infected are usually non-migratory birds,

Where the identity of the species of infected wild bird is known (many cases do not identify the infected bird species), in most cases the species involved are not long distance migrants, but species which move relatively short distances, and are mainly associated with dispersal movements (not migrations) associated with weather. (BOU 2006)

GRAIN adds to the evidence that wild birds are not the source of the problem. In citing FAO (Food and Agriculture Organization, UN) experts, officials from GRAIN assert,

To date, extensive testing of clinically normal migratory birds in the infected countries has not produced any positive results for H5N1 so far. Nearly all wild birds that have tested positive for the disease were dead and, in most cases, found near to outbreaks in domestic poultry. Even with the current cases of H5N1 in wild birds in Europe, experts agree that these birds probably contracted the virus in the Black Sea region, where H5N1 is well-established in poultry, and died while heading westward to escape the unusually cold conditions in the area. (GRAIN 2006, 5)

The BOU also states that risk of infection is 'extremely low' and reiterates that human cases of H5N1 occurred in people working with infected poultry.

Given that the direction of infection seems to flow from the industrial system into the wild bird population, it is important for decision-makers to be well-informed. Misinformation can prompt public officials to adopt measures that run contrary to public well-being including: 1) the culling of wild birds as this destroys vital genetic resources that may be needed to resist the spread of diseases; and, 2) The destruction of wetlands. These too need to be conserved as sites of biologically diverse habitat (BOU 2006). According to the BOU there is a strong need for informed groups such as conservationists, scientists and veterinarians to clarify the issues, protect wild stocks and control the spread of the disease. It is suggested that these experts need to work closely with the media,

> ...to give policy makers, stakeholders and the general public more balanced information on real levels of risk and appropriate responses" and to avoid, "'mixed messages' that public can sometimes receive regarding risk. Whilst there is an assessed low risk of human infection by H5N1 from wild birds, photographs of professional staff dressed in comprehensive personal protective equipment can suggest otherwise...All agreed that it was most important for conservation bodies to avoid the development of a public culture of fear associated with wild birds and to encourage similar attitudes in other relevant professions. (BOU 2006)

Other recommendations to avoid a flu crisis include the creation of surveillance, data collection and collaborative information sharing relationships that engage multiple stakeholders from a wide range of jurisdictions including international, national and local governments, agencies and NGOS. The creation of an early warning system is also needed to flag new cases and contain the spread if and when it happens.

The case of Laos offers valuable insights into dealing with bird flu and the potential for alterative poultry management systems. In Laos two separate poultry systems have emerged. Unlike other countries where there is crossover between veterinary services, hatchlings, markets and feed for both domestic and export poultry markets, Laos has two streams. A USDA report cited in GRAIN (2006) documents the low incidence of bird flu in Laos and the notably small crossover between small, domestic focused poultry operations and large, export-focused producers,

> A total of 45 outbreaks were confirmed, with 42 of these occurring on commercial enterprises (broiler and layer farms), 38 of these in Vientiane, the capitol and primary city of Laos. Another five outbreaks were found in Savannakhet Province (on one layer farm and in smallholder flocks) and another two in Champasak Province (on layer farms). Smallholders who found avian influenza in their flocks were located nearby commercial operations suffering the disease. (GRAIN 2006, 9)

As the GRAIN report concludes the low rates of bird flu exist despite proximity to highly infected countries such as Viet Nam and Thailand due to the separation of low risk, backyard small-scale operations from high risk, large scale industrial farms.

The case of avian flu raises two important points. First, disease is not easily contained particularly given the global flows of people and food products (Harris Ali and Keil 2006). In the case of the Suffolk turkeys, although the report stated that the occurrence of H5N1 was 'rare', the disease spread within the flock despite high levels of bio-security measures in place at the site. This points to the high susceptibility of domestic birds and the vulnerability of the industrial system. Second, the industrial food system is able to promote its own self-interest even in the face of a global pandemic. The industrial poultry reaction is to ramp up bio-security measures. A parallel response has been to eliminate small backyard flocks despite evidence that factory farms pose the highest threat. In fact the genetic diversity contained in wild and heirloom breeds may be the only way to counter the impact of a pandemic if it does strike as the diversity would act like a circuit breaker in stopping the spread of disease.

Returning to the beginning of the chapter and the engagement with the concept of food as actor, the translation of the circumstances surrounding the disease is a critical point. Translation nodes such as the media and the globalized food industry shape the way that politicians and the public react,

> Now, with the H5N1 outbreaks, people are dying because of this industry, and the problem will never go away as long as factory farming continues to expand and operate without accountability. Bird flu is yet another of the scandals that have played out time and again with other sectors of the transnational food industry, from mad cow disease to Star Link maize. It is simply shameful that the poultry industry is trying to spin it into another growth opportunity on the back of small farmers. (GRAIN 2006, 18)

Technology, Science and the Management of Public Risk

In Chapter 1 we explored risk as both reflexive and socially constructed. In the context of disease and food, several actors contribute to the construction of the way risk is perceived and managed. In the context of disease and related health scares, one strategy on the part of the food industry is to offer high technology solutions to resolve existing and looming crises. Ironically, the result is the layering of one complex technology on top of another as various branches of the food industry use new technology to resolve a problem created by an old technology. The industry responds to weeds that persist in mono-crop agricultural systems with plants genetically engineered to tolerate the application of herbicides. In the case of avian flu, one response is to develop GE poultry engineered to be resistant to influenza. A scientist from Cambridge University cited in The Times, London report that,

> Researchers in the UK are pursuing transgenic bird flu-resistant chickens. "Once we have regulatory approval, we believe it will only take between four and five years to breed enough chickens to replace the entire world population," said Laurence Tiley, Professor of Molecular Virology at Cambridge University. (Henderson 2005)

Irradiation is another technology fix that is heavily promoted by industrial food proponents as a means to deal with food contamination such as salmonella and trichinosis. By exposing food to ionizing radiation, shelf life can be increased, ripening and sprouting delayed and pathogens deactivated (Scott and Elliott 2006). Interestingly though, the history of irradiation in the US demonstrates the power of advocates to stall a technology through the successful lobbying of policy-makers and consumers. Despite approval by US federal agencies, the WHO, the FDA and several US scientific and medical organizations there are only 30 irradiation facilities in the US. These sites are used primarily to sterilize medical equipment although some treat food. The USDA supports food irradiation as a more benign alternative to chemical fumigation for spices and proclaims its use for food fed to patients with compromised immune systems. Invoking high tech and patriotism, they also explain that, "American astronauts have eaten irradiated foods in space since the early 1970s" (USDA 2000, 1). Despite this enthusiasm, consumer backlash linked to fears of terrorism aimed at irradiation facilities and the risks associated with having this technology resulted in a strong NIMBY response (Ten Eyck 1999). In this context, the media is an important interpreter and acts as a mediator between experts and the public. Glassner (1999) in line with thinking by Giddens on the environment of chance and risk argues that much of this risky environment,

> ...has been created by the popular media that focus on hazards that often are more sensational than pervasive or factual. Regardless of whether risk is real or a figment of media presentation, however, most of us are left with little choice than to rely on the advice of experts to navigate through these seemingly hazardous times. (Glassner 1999, 822)

The Alar incident was a case in point. As Carolan (2006) reminds us, trust and power are critical to the determination and communication about risk,

> To speak of risks is to speak of an epistemic void – for again, if we knew the consequences of object X or action Y it would not be a statement of risk but a statement of factual certainty. So we must look toward our social relations of trust to give these risk statements meaning and fill this epistemic void. Such trust provides us with a needed compass by allowing us to navigate through the numerous risk statements that flood our lifeworlds daily. It allows us to separate 'truths' from 'untruths', and in so doing provides us with a means through which to make sense of the complex, epistemologically distant, world around us. (Carolan 2006, 247–248)

As we have seen in this chapter, food has an active role in defining relations in the food system. As a vector for disease, it is involved in the production chain in a way that is unlike other commodities. Different actors along the food chain translate the power of food in different ways. The food industry has used food scares as one rationale to increase the level of technology and control in the food chain. Experts and the media interpret food scare information for consumers and can turn public opinion in one direction or another. In the end it is the power to persuade and be trusted by consumers that determines perceptions about food and risk. Ambivalence about the food system leads to an explosion of anxieties for consumers and provokes a search for alternatives to the IFS. Consumers want to be confident that what they are eating is healthy, free from chemicals and disease. However, the public is developing a growing conviction that the complex IFS has evolved with so many interacting and conflicting priorities that it may create more variables than can be controlled or regulated meaningfully. Perhaps, in our quest for better food we have made the system worse. In the next chapter we explore food fears from the consumer perspective through the lens of organic food as a choice to get back power from the IFS. As we see, current conditions in the US surrounding terrorism and food add to fears in the aftermath of 9-11.

Chapter 6

Eating Organic in an Age of Insecurity

Betsy Donald and Alison Blay-Palmer

For the life of me, I cannot understand why the terrorists have not attacked our food supply, because it is so easy to do. (US Health and Human Services Secretary T. Thompson, 2004 in Hulme 2005)

This chapter delves deeper into the complexities of the industrial food system to understand better our fearful relationship with food. Building upon the ideas established in previous chapters about the increasingly uncontrollable effects of the industrial food system, we examine the construction of societal fears since 9-11, the advent of food sovereignty and bio-terrorism, and the reaction of consumers in the wake of these new pressures. We interpret food fear in the context of the changing role of the state in dealing with deeper societal insecurity. We argue that with the retreat of the state in many social realms and the concurrent heightened security interventions, many people have adopted auto-regulatory strategies in their attempt to manage broader societal risks (Beck 1992; Foucault 1977). We point to the growth of organic food in North America as an excellent example of this newer type of self-regulatory behaviour. We base our argument on empirical results, but also upon insights from literature in political sociology and on political geographies of North American food industrialization. In particular, we revisit Beck's (1992) conception of the 'risk society' introduced in Chapter 1. In this chapter we stress the importance of risk management at multiple scales and identify iteration between the individual scale to the global.

To some extent, an individual's response to food fear is embodied in their food consumption (Halkier 2004). The individualized, reflexive response provides an important context for understanding eating organics as a form of 'self-regulatory' modernity (Beck 1992). In the context of food fears we are especially intrigued with politics at the scale of the body where personal and state power relations are negotiated (Foucault 1977). We are interested in the ways that the socialization of food fear conforms to Foucault's idea of governmentality so that "this type of governing is focused on normalizing and disciplining people's bodily practices at a micro-level" (Hewitt 1991, 228–230 in Halkier 2004). Building upon Foucault's concept of governmentality and bio-politics, Bobrow-Strain (2006) makes a compelling link between: 1) the industrialization of North American food in the early 1900s, 2) food safety activism, and 3) body politics. A provocative idea from his paper on the history of bread in North America is that "food safety activism is ill-suited, and perhaps counterproductive, to the challenges of creating a radically democratic food system" (Bobrow-Strain 2005, 6). In the context of contemporary food fears this suggests two things about the importance of scale when trying to address food safety issues in a meaningful way. First, the contradictions of containing germs and regulating food by national and international regulators may be in conflict with the creation of a scaled-down, more locally embedded food system. This leads us to ask whether a more personal food production-consumption relation may in fact be relevant for achieving a safe food system. Second, we suggest that the surge of interest in organic represents a contested space between the industrial food system and a new more sustainable one.

In this emerging food reality there are forces emanating from multiple scales, from the body, to the home, community, region, nation and globe (Bell and Valentine 1997). Bell and Valentine explore tensions at the intersection of culture and consumption as they emphasize the messiness of food and the 'spaces between' scales (Bell and Valentine 1997, 12). So for example, they discuss 'glocalization' and the way that food is constantly reinvented through time and space. They use the example of pasta's migration from China to Italy and its assimilation into Italian culture so that pasta as a version of Chinese noodles now sits at the centre of Italian cuisine. Pasta becomes a symbol for an early version of McDonaldization as food cultures converge (Ritzer 2006). At the more personal end of Bell and Valentine's analysis are the individuals with food disorders who deal with the enactment of food consumption at the individual scale – for example, anorexia. This telescoping in and out from one scale to another informs this chapter as we move from self-regulation and organics as an individual response to complex IFSs through to the national and global effects of 9-11 where the threat of bio-terrorism has accelerated the adoption of computerized, radio food surveillance systems that can track food to its source in four hours. Oscillating between different scales helps us to understand the complexities of the industrial food system and food safety from multiple perspectives.

The current emphasis on military preparedness and homeland security that emerged from 9-11 contorts all facets of our everyday lives including our food spaces. In this excerpt from a paper prepared under the guidance of Colonel Coleman at the US Industrial College of the Armed Forces National Defense University, it becomes very clear that the scale of the industrial food system makes it vulnerable to attack,

> Since September 11, 2001, food security has become an increasingly important national security issue. Previously, the possibility of intentional contamination of the national food supply was considered so remote as to be insignificant. Today, however, many are concerned with the risk of deliberate attacks. Recognizing this risk, President Bush added the agriculture and food industries to the list of critical infrastructure sectors needing protection from terrorist attack. Since the 9/11 attacks, much work has been done to improve security of US food supplies during the preparation, processing, packaging, and serving phases. However, keeping food secure while it moves from point to point in the system remains a significant concern and is an area where further security improvements are required. (Coleman et al. 2003, 5)

And so the everyday act of consuming food is now linked explicitly to the 'War on Terror'.

Given the military's renewed interest in food and war, we make connections between eating organic as a self-regulatory act and a state regime that emphasizes eating healthier food as part of broader state social and military policies. We see that in the same way that people eat organic food to 'get control' over what they are putting into themselves, national food surveillance systems are being refined to 'get control' over an increasingly complex industrial food system. Both actions in the end reflect a need to deal with food fears created by the IFS at different scales. However, before reflecting on issues in the national domain, we report more about our empirical study into the expanding organic industry in the Toronto city-region and the body-politics of eating organic.

Organic Food as a Self-regulating System

Organic consumption in North America has grown at a rate of 20% per year for over ten years (Willer and Youssefi 2007) and was valued at $10.4 billion in 2003 (Dimitri and Oberholtzer, 2005). In 2005, the global market was worth over 25 billion euros and is expected to exceed 30 billion euros in 2006 (Willer and Youssefi 2007). Our research into the Toronto organic industry points out that some of this growth is linked to producer and consumer fears about the risks involved in eating from the industrial food system (Blay-Palmer and Donald 2006). This chapter explores these fears in the context of terrorism, control, personal well-being and trust-based food systems. As a food processor told us, "[W]e are worried about terrorists. This is nothing to worry about in comparison to what we are doing to ourselves".

Consumers who eat locally produced organics see this food alternative as a way to ensure their food safety as well as a way to *control* the quality of the food they eat through shorter, trust-based food networks. These localized or certified food networks provide the consumer with the means to regain trust in the food they eat (Goodman 2003; Hinrichs 2003; Ilbery and Kneafsey 2000; Murdoch et al. 2000). In work by Ilbery and Kneafsey (2000) on the role of self-regulation by producers in alternative, short food supply chains,

> It emerges that, despite new regulatory frameworks and consumer concerns, producers usually define quality in terms of product specification and attraction rather than through social certification schemes or association with region of origin. Food quality, however defined by producers, is essentially self-regulated and constructed within the context of maintaining stable relationships between producers and buyers. (Ilbery and Kneafsey, 2000, 217)

In interviews with retailers, processors, producers, distributors and consumers, we discovered that organic innovators were motivated by their own personal fears and related personal health issues. One food processor described some of the health issues that led her to explore alternative eating options,

> I had dietary issues from eating conventional food. My passivity regarding my food purchases led to me overwhelming [my] body with certain food groups (gluten intolerant), so I looked around, frequented [farmers'] markets where people advised me about the food I ate. Then I bought organic strawberries and ate them. I had no reaction. I bought an organic orange, and I was able to eat it. I had no pain despite having been unable to eat oranges since my childhood. Because the conventional food industry is controlled by some very money-driven people who are looking away from the fact that they don't know what the chemicals etc. will do to you and I. (Food processor, February 2003)

She then explains that the conventional food system let her down and is part of a larger societal failure,

> It annoys me that organic food has to be labeled but poisoned [mainstream] food doesn't. Parts of processing that uses chemicals doesn't [sic] have to be labeled. Children are sick. How much is being caused by the food we eat? (Food processor, February 2003)

Finally she explains that engaging with the organic food system and recovering her health led her to want to spread this well-being to others,

> I ask myself, what do I do to give back to the community? I want to inform, to advocate on behalf of what I believe in after the tech bubble broke, 911, collapse of Enron, Anderson everything changed. (Food processor, February 2003)

These personal as well as societal dilemmas – which were linked to human health concerns and a general environmental decay – were also raised by one retail manager,

> ...our customers are concerned about health and the environment and how we fit in it. They see food both as fuel and a source of enjoyment. Our customers want artificial-free food...We have a commitment to environmental ecology. (Retailer, February 2003)

The IFS and its emphasis on mass production has hollowed out the food system and created a crisis of confidence for food consumers, "[b]ig business and the environment are damaged in terms of the pollution they create. For example, soils get pummeled because of demands from conventional agriculture" (Food processor, February 2003). The quality of food itself has also been compromised. In assessing the nutrient content of food, experts are finding that food produced in the industrial food system is less nutritious now than in the 1940s. This means we need to eat more food now to get the same nutrients that were available from food grown in the 1940s (Pollan 2006; Pawlick 2006). There is also evidence of other personal health issues associated with consuming conventional food, "[t]he general health of the population exerts an influence, especially the occurrence of unusual health problems such as allergies and pre-pubescence [sic]. There are food-related environmental problems as well such as GMOs" (Food processor, February 2003).

A related personal health risk is food allergies. Peanut allergies in particular provide relevant insights into one aspect of the fear-risk-food discussion. It is well documented that peanut allergies can be fatal for some people. As well, the number of people affected by peanut allergies is on the rise. It was estimated in a clinical study involving over 4300 children in Montreal that on the order of 1.5% (confidence interval 95%, 1.16–1.95 range) of children have peanut allergies (Kagan et al. 2003). Research in the UK conducted by the David Hide Asthma and Allergy Research Centre determined the number of children who tested positive for a peanut allergy was three times the level in 1989 (1.1% in 1989 versus 3.3% in 2002) (BBC 2002). This is consistent with findings in other research studies in the US. The threat of peanuts is so severe for some children that most schools and childcare facilities now prohibit the use of peanuts. The fear of a severe reaction forces parents and institutions to take extreme measures to protect allergic children. One problem in trying to enforce these new standards is the pervasive presence of peanuts in the industrial food system. The UK Food Standards Agency found trace amounts of peanuts in 56% of processed food (BBC 2002). In this light, keeping allergic children safe is a huge challenge. Peanut allergies point to two issues of particular interest. First, the complexity of the supply chain prevents people from knowing what is in their food, creating a serious and very real threat to people with peanut allergies. The second interesting feature is the moral regulation questions that are raised – that is, "moral in that they project a vision of a carefully regulated" and consequently constructed social contexts where persons of authority are responsible for "an expanding range of risks to the physical, sexual and moral well-being" of the general population (Rousa and Hunt 2004, 826). As we move increasingly in the direction of regulating different aspects of our society to mitigate and control risk, we lose control over the details that are relevant for the individual. This creates a personal problem that is difficult for a global industrial food system to address. So we have the majority of processed food products that warn 'This product may contain peanuts' and people who have very limited food and lifestyle choices as a result of their food allergy. It is also worth considering that there has been a parallel loss of choice for people without allergies that effects their dietary options – this point recurs in the context of school meal programs discussed in Chapter 8. Food allergies point to another facet of the difficulty in managing an integrated IFS.

From one retailer's perspective, the desire on the part of consumers to deal with these societal food challenges by changing their 'health-destiny' creates opportunity for the future, "[c]oncern about GMOs, pesticides, chemicals and additives to food will create new niche markets" (Retailer, February 2003), and "[h]ealth crises are going to push us (for example, Mad Cow) as people realize that they need safe food and that they have to support sustainable agriculture to get it" (Retailer, February 2003). The destruction of food and environmental quality creates a direct opportunity to increase business for organic producers, processors and retailers. As this farmer/ processor explains the family-owned beef farm and retail meat outlet are physically separate as a food safety precaution,

> We wanted to distance the farm from food processing due to hoof and mouth disease because of bio-security issues. We strive to be very customer service oriented so they are very loyal. (Farmer, processor, April 2003)

The increased interest in quality is based to a large extent on trust, "there is a loyalty. We are a part of the community. We are recognized as a leader, an innovator". (Retailer, April 2003). As one chef put it, "[l]oyalty is very important, when there is a crisis (e.g. BSE, SARS) it has an immediate impact on the industry" (Chef, June 2003). Proximate, trust-based relationships are needed to confirm that, "organic is real and organic is the real deal, public perception is really the key to the value of the industry" (Distributor, February 2003).

Our empirical work illustrates the attraction of eating local organic as a space where consumers can control one aspect of their lives. By deciding on what they are going to eat, consumers bypass the political-economy of industrial food systems and opt to consume in a context where they have assurances of safety and quality. These assurances sometimes come from confidence in certified product, but more deeply from trust-based, face-to-face interactions with food producers and processors. In this way they bypass the threats inherent to the IFS.

Contradictory Nature of the Organic Food System

Not all is crystal-clear concerning this self-regulatory system, however. As Guthman (2004 a, b) has argued, much of the organic system itself – in the California context at least – has been co-opted by the industrial agro-food industrial system (see also Walker 2004). The entry of large companies into the organic realm has generated market confusion over whether or not people who are eating organics are in fact supporting a more sustainable 'alternative' to 'actually existing forms' of neo-liberal industrialized food systems. Part of the self-regulatory act of consuming organics, as we discovered in our research, was the belief by many that they were actually participating in a local, small-scale food revolution.

Yet the paradigmatic potential of this alleged revolution is somewhat overstated when one examines the more recent consolidation of the organic industry especially from the retailer and distribution perspective. According to a company official, Sun-Opta currently controls 90% of the distribution of organic foods in Canada. Sun-Opta Inc. started as a privately owned company interested in alternative energy and organic food. In the course of its rapid growth it took over ProOrganics, the largest Canadian organic distribution company. ProOrganics had its roots in the 1970s food movement on the west coast of Vancouver. As one of the original alternative food pioneers in Canada, it had built extensive national networks in the organic community and had developed very deep social capital. It was seen as a cornerstone of the alternative food movement. Sun-Opta consolidated ProOrganics with three other distribution companies to create the most extensive organic and kosher distribution network in Canada. With divisions in western, central and eastern Canada, its reach is truly national in scope. Sun-Opta is now a rapidly growing, publicly traded, vertically integrated international company with expertise in sourcing, processing, packaging and worldwide distribution of natural, organic and specialty foods.

The Sun-Opta Grants and Foods division of the company offers a variety of packaged food products for retail and food service use. The vertically integrated nature of their organization enables them to control product quality and consistency from seed selection through to processing and packaging. According to a company official Sun-Opta produces "private label" and other health products for a number of well-known retailers and consumer food companies (e.g., Costco and Sam's Club) as well as processing their own products, specializing in producing healthy, nutritious and organic snacks and food service items to facilities across North American, including the military.

The Military and the Body Politics of Diet

The military's interest in organics is a fascinating one, as it demonstrates how eating organic in an age of insecurity is part of a new state regime that emphasises eating organic as part of a broader state social and military policy. It exemplifies the conflicts that permeate the IFS. Over the years the military has been preoccupied with quantity and quality. At different times, this led to ramping up volume, and process efficiency and reinforces industrial food production. Now, having helped to construct the system, they are trying to find ways to eat differently.

Contemporary media stories on obesity frame the debate not only in terms of individual responsibility, but also in terms of national security. Obesity – just like malnutrition did during previous wars – threatens national security, compromising physical readiness and the "manpower" of military forces. Long before obesity was on civilian radar screens as a health issue, the US military identified the problem and responded with an aggressive health and diet campaign (Vogel, 1992). The Department of National Defense is a major funder of obesity-related research and military bases now offer organic options to their soldiers and families as part of healthier living initiatives (Norris, 2006).

The military's current concern with the health of the population reflects a deep historical interest in the role of diet as it relates to military preparedness (Shilling, 2003). Morgan (2006) has noted that in the UK, war has been just as fundamental as welfare in shaping food policy as far back as 1893 when the soldier's poor physique weakened the Boer War effort. During the First and Second World Wars, food shortages in the UK also helped shape a food culture based on a 'food as fuel' ethos. And similar to the UK example, individual dietary politics in the US have become intimately connected to national politics. War is not fought on behalf of a sovereign that must be defended, but rather on behalf of a society that must be defended. In other words, the logic of war is about protecting a population from internal and external threats to its health (Lemke 2001). So that in discussing the body politics of diet through bread, Bobrow-Strain (2006) points to the ways in which eating habits are a field of power relations made up by expert knowledge, surveillance and discipline. WWII, for instance, was a time of great dietary surveillance, discipline and experimentation. In the realm of food production, bread emerged as a source of national pride, competence and a vehicle to fortify military troops. Between 1941 and 1943, the War Food Office initiated a series of Federal laws that mandated increasingly stringent levels of nutrtitional fortification for all white bread manufactured in the US. By 1942, 80% of American's white bread was being fortified with thiamine, dry milk, niacin and iron. Although mandatory fortification ended in 1944, it still continues in practice to this day (Bobrow-Strain, 2006).

The pivotal role of food in WWII was summarized in a 1941 Time Magazine article as,

> Food, the most important single element in morale, is a crucial factor in World War II, and the struggle for it was one of the major battles last week. The plantings and the harvests of 1941 and 1942, if their power were understood and they were properly geared into the major strategy of the war, might be the determining factor. Certainly the nation which won the Battle of Food would sit at the head of the peace-conference table.

The awakening to the importance of nutrition and vitamins continued into WWII as the US government shifted from food quantity as 'Food Will Win the War' to food quality as 'Vitamins Will Win the War' (Levenstein 1993, 64). Emerging from the Depression of the 1930s, the Surgeon General estimated that one third of Americans were mal-nourished. In a Time Magazine article describing the need to ramp up US agricultural production to 'feed Britain' and 'starve Europe' concerns about over-production are met with the following comment that reveals both the power of the US government to establish prices as the prime food suppliers to the war effort; it also reflects the depth of undernourishment in the US,

> Perkins [Assistant to the US Secretary of Agriculture] holds the whip over prices, through the power of his surplus-buying. Any time he quits buying, he can wreck the market. This power kills food speculation. Outside of cotton, wheat and tobacco, they consider that there is no surplus problem except storage space – because they can always dump extra purchases into the daily free school lunches of 9,000,000 undernourished U.S. children, or into the tremendous maw of the food-stamp program, feeding relief workers. (Time Magazine 1941)

Under directives from President Roosevelt, a set of Recommended Daily Allowances (RDA) was established in 1941 as a way to measure daily nutritional requirements. The RDA provided, "an official-looking yardstick for judging how well everyone from factory workers in Chicago to GIs in New Guinea were being fed" (Levenstein 1993, 66). While men in the military received extra doses of vitamin A and B complex, the merits of supplementing factory worker nutrition was debated. Factory worker productivity was seen as critical to a successful war outcome but in the end, the decision was made to educate instead of supplement. Public education was deemed to be the most effective, and less authoritarian, way of getting all US citizens to eat more balanced meals (Levenstein 1993).

As a way to clarify what foods should be eaten, the US Nutrition Division developed a seal of approval that included the phrase the 'US Needs US Strong, Eat Nutritional Food'. This label was embraced by producers and processors alike and was included in many brand-name and association advertisements. Kellogg's offered bonuses to restaurant staff if they promoted the Kellogg's cereal and got "customers to 'step up' to its whole grain or 'restored' cereals" (Levenstein 1993, 75). Advertisers during the war also linked their products to war efforts,

> Ads for "Nutritious" Netstle's cocoa said chocolate had been selected by the U.S. Army for its Type D emergency rations "because it is a concentrated energizing food". Candy manufacturers played other notes from the energy theme, sending the schools charts showing how sugar was used in the "combat and emergency foods of U.S. fighting forces". They even persuaded the first lady, Eleanor Roosevelt, and the army quarter-master general to appear in a nationwide broadcast supporting their campaign to have Americans send tons of confections to loved ones in the service to "fortify energy" and "boost morale". (Levenstein 1993, 76)

In this way food labeling veered in a new direction as endorsements from experts and approved labels became the standard for selling a product for its 'nutritional' value. As WWII continued, associations between food, patriotism, war and health were reinforced. For example, Florida grapefruit was described as the "Commando Fruit" high in "Victory Vitamin C" (Levenstein 1993, 76). The war introduced another innovation for food – as food rationing was introduced in 1943, farmers secured $50 million for the procurement of abundant local food for mostly rural school children. In all then WWII introduced pivotal innovations to nutrition, food marketing and procurement.

Linking Military State Objectives to Food Industrialization: Managing Risk in the Contemporary Agro-food Context

The public's desire to manage risk through the act of eating has taken a different trajectory over the last five decades as food scares have become more and more connected with industrial agri-food practices. The petro-chemical agricultural industry can trace its roots back to the military and the development of wartime technologies. The most notable example is DDT. A relative of chemicals and technologies developed in WWI, DDT was widely used in WWII for insect control in the fight against typhus and malaria epidemics (Walker, 2004). Its success was so great that a news magazine reported in 1944 that, "[p]roduction has multiplied 350-fold in the last year [1943-44]; four manufacturers are now turning out about 350,000 pounds a month-all for the Army" (Time Magazine 1944). And so DDT and other "pesticides became a basic part of the petro-farming package at the middle of the 20th century" (Walker, 2004: 185). Although the dangers of DDT were made clear by the early 1960s (e.g. Rachel Carson's Silent Spring 1962) the groundwork was established for the wide scale acceptance of chemicals as part of a progressive farming regime, facilitating the Green Revolution. The trilogy of high yield varieties, chemical applications as fertilizers, pesticides and herbicides and the use of irrigation doubled crop yields every decade since their inception in the mid-1950s into the 1980s. But for many producers and consumers, the public health and environmental costs of producing this type of chemically infused system has simply been too great.

For some, the relentless industrialization of food over the last half-century has created new 'invisible' and pervasive threats. As we discussed in Chapter 5, food scare events such as BSE-contaminated meat and GE food products have lead to fears that are both geographically pervasive but also invisible through the traveling of pollution and disease in keeping with Beck's (1992) notion of invisible risks and Whatmore's (2002) of hybrid geographies:

> The metabolic impressions that the flesh of others imparts to our own is an enduring axiom of social relations with the non-human world and the porosity of the imagined borders which mark 'us' off from 'it'. The potency of this vector of inter-corporeality seems to grow as the moments and spaces of cultivating and eating, animal and meat, plant and fruit, become ever more convoluted. The troubling spectres of fleshy mutability that haunt the shadowy regions between field and plate mass with particular intensity in the event of 'food scares'. Such events are emblematic of the threadbare fabric of trust (dis)connecting industrial food production and consumption as we enter the twenty-first century. (Whatmore, 2002, 120)

Organic agriculture offers a viable alternative to an anxious public.

The Power of Alternatives

What does eating organic mean, then, in our age of insecurity? In the current era
of invisible as well as more tangible food and even wartime threats, eating organic
becomes a symbol and a tool for people to recapture their power. The act of eating
is one occasion where people can decide what will be incorporated into their bodies,
thus transforming the individual's role as a taker in a modern society, to a participant
in a more reflexive-modern 'risk' society (Beck, 1992). We thus see eating organic
today as an example of how a self-regulatory act – or what others have called a
very intimate body politic (Whatmore 2002; Goodman 1999; Foucault 1977) – can
take on great meaning during wartime (cf. Bobrow-Strain 2006). As we experience
this period in history overshadowed by terror and war, we must ask ourselves the
extent to which self-regulatory behaviour such as eating in an 'alternative' way (e.g.,
through consuming organic product) is in fact in opposition to – or fundamentally
part of – the current state regime, comprising a part of a renewed military-industrial
complex. In the US the response to perceived bio-terrorism threats to food security
has translated into increased regulation over food and the associated technological
control and surveillance that now exists in the industrial food system. By 2006,

> ...the Food and Drug Administration finalized section 306 of the Public Health security
> and Bioterrorism Preparedness and Response Act of 2002, or "Track and Trace," that
> requires most every business in the U.S. food supply chain to keep detailed records on
> receipt and shipment of goods – where they come from, who they've been sent to, lot
> numbers, and more – and to be able to supply that information four to eight hours after it's
> requested. Other rules in the act require companies to register food facilities and provide
> advanced notice of food shipments coming from abroad. (Hulme 2005, 40)

The 'granular' tracing capabilities that are now standard technology in the US food
industry allow processors and distributors to track food. Hulme provides the example
of the Bama Company to illustrate the necessary level of sophistication and control
as the food industry attempts to ensure safe food,

> The Bama Cos., which makes pies, cookies, and frozen-dough products for companies
> such as McDonald's and Taco Bell, uses a bar-code seal that records the serial numbers on
> rail shipments of the corn syrup it uses in its products. Track and Trace "is a big deal right
> now", CFO Bill Chew says, necessitating a big change in how the company keeps a log of
> the ingredients it uses. (Hulme 2005, 40)

The Track and Trace technology helps to monitor the vast quantities of processed
food that get shipped around the country,

> Not too long ago, each time a rail shipment of corn syrup arrived, a worker had to climb on
> top of the rail tanker car to manually record a unique seal-identification number and then
> input the information into Bama's tracking system. If the seal was tampered with, broken,
> or improperly recorded, the shipment could be rejected and sent for safety testing, Chew
> says. Such lost shipments can cost the company $25,000. Bama records about 60,000
> seals each year, and having workers run from railcars back to the office to input the data
> was expensive, with too much room for errors, Chew says. (Hulme 2005, 41)

Dealing with this quantity of food processing requires sophisticated technology,

> As a result, earlier this year the company installed a system called SealTrac, co-developed by mobile-computing device maker Psion Teklogix Inc. and mobile application maker Anyware Mobile Solutions, to better track safety seals. Workers carry a Psion handheld computer with Anyware software to scan bar-code information from safety seals on railcars, as well as on hoses used to siphon out the syrup and on the tanks where the syrup is stored. The data is then wirelessly transmitted to a central database to immediately record that a shipment is available for use in production. "You don't have to run numbers to see that if you can automate this process and automatically scan it, it eliminates a huge opportunity for error and cuts labor costs", Chew says. (Hulme 2005, 41)

These new systems are very high tech and would be beyond the means of small-scale food processors. Effectively then the new Track and Trace requirements privilege the IFS. At the same time, Track and Trace scales up production and reinforces the circumstances that create the original threat. The vulnerability in the industrial food system lies in its vast dimensions and complexity. Smaller food systems, for example a farmers' market, do not have the same reach. If a batch of spinach from a farmers' market is contaminated then the scale remains local. If a company with national or global reach has a contamination problem, the effect is much different. For example in October 2007, ConAgra recalled all of its beef, chicken and turkey pot pies from stores that sell their lines Albertson's, Hill Country Fare, Food Lion, Great Value (sold at Wal-Mart stores), Kirkwood, Kroger, Meijer and Western Family as 165 people in 31 states contracted salmonella found in the pies. A product recall on this scale is more complicated than tracking down people on a more local level. The implications are also – literally – more far-reaching. With the retreat of the state from social welfare and its heavier hand on the more repressive side, many people have developed auto-regulatory behaviour in their attempt to manage broader societal risks and fears.

Organic food consumption is expanding rapidly throughout the cities of North America and in particular amongst the highly-educated, the environmentally conscious and the politically active (Blay-Palmer and Donald 2006). The act of consuming raises ethical questions that can inform personal decisions about food systems. Garbriel and Lang (2006) challenge us to consider the interaction between global economies and personal capacities,

> Inequalities among consumers [that] are already sharp, leaving substantial numbers of them window-shopping with only restricted opportunities to make a purchase and many, in the developing countries, without even windows to window-shop. (Gabriel and Lang 2006, 193)

And then, there are the emotional, cultural, political and personal spaces where for example, western over-consumption and Muslim fundamentalism collide to feed current international tensions and insecurity.

Problems of inequity in the developed world also need to be acknowledged. Canada – considered by many to be one of the best countries to live anywhere in the world – reports food insecurity at 15%. In the context of organic food, this provokes questions about people who cannot afford the generally higher prices of eating organic, even though they themselves may be anxious about the long-term health and nutritional costs of eating conventional food? Ironically, it is the large, multinational corporate institution that has responded to their call. Retailers like Wal-Mart have introduced organic food into their product line-up. While Wal-Mart claims it will democratize organic food and make it more affordable for the masses, others argue that the retailer's push into organics will hurt small-scale local organic farmers and lower environmental standards for the production of organic food. Nutritionists like Marion Nestle have also questioned the nutritional benefits of Wal-Mart's involvement, arguing that many of these newer mass-industrialized, highly processed organic products lack nutritional content (Warner 2006).

In food terms, the question becomes: who now shapes broader social and health policy on issues of food nutrition and health? With the retreat of the state on many social dimensions, do we now look increasingly to the military-industrial complex for our direction in these matters? Will the governments' efforts to protect and improve the American population from internal and external threats to its health ultimately reshape public health policy on food production and consumption? Or will this once active area of social policy be left up to the individual and corporate response? In posing these questions and unpacking issues from multiple scales, the increasingly complex dimensions of industrial food production resurface. But we are again left with the sense that complexity has hollowed out the industrial food system leaving it more vulnerable and increasingly difficult to secure. In a more hopeful turn, we explore food small and medium sized food processing companies in the next chapter. They provide a starting point for constructing more sustainable, community-based food systems.

Chapter 7

Manufacturing Food Fear

Alison Blay-Palmer and Betsy Donald

Seeking to prune costs most large manufacturers had gradually engineered natural ingredients out of food, replacing them with cheaper, artificial flavours and preservatives that could ensure longer shelf life and were easier to handle. (Kingston 1994, 119)

In Chapter 7 we focus on issues of food safety and how they mould the alternative food industry. Essentially we argue that by reacting incrementally to fairly specific food safety issues created by the industrialization of food processing, the 'alternative' plays into the hands of the industrial. Our research and analysis from empirical work in Toronto confirms that the expression of 'alternative' in specific, codified terms enables the cooption of alternative market segments by dominant agro-food players. As a result, the alternative food system is repeatedly undermined as industrial food cannibalizes bits of the alternative and absorbs them as branches of the mainstream food. The role of players such as Wal-Mart and McDonald's in this new food world will be discussed.

In this chapter we present the case that AFSs are weakened in part as a result of socio-cultural factors. Specifically we explore: 1. the response of the alternative food industry to food safety issues; and, 2. food scares as one phenomenon that define AFSs. Of particular interest for this chapter is the relatively unexplored middle space between the producer and the consumer; and so our focus is on food processors and retailers as the intermediary nexus in the alternative food system. Although we have a developing and thoughtful literature on the production and consumption ends of the food chain, little is known about alternative food retailers, and even less about processors. As processors and retailers interpret, mediate and translate the food that passes from producer to consumer, they are a critical but poorly understood point in the food chain. Further, the research that has been done on these intermediate spaces tends to focus on rural-based processing and/or retail links (Ilbery and Maye 2005b; Ilbery et al. 2005; Hinrichs 2000; Holloway and Kneafsey 2000). This chapter takes us into an urban setting by exploring alternative food systems in Toronto, Canada. By addressing the opportunities and challenges that emerge for alternative food systems in large cities, we begin to address the urban and retailer/processor gap in the alternative food literature (Donald and Blay-Palmer 2006). Accordingly, we investigate urban food systems through the lens of food safety and the way that North American consumer fears are translated into products and shelf-space by processors and retailers. The chapter builds on Goodman's conceptualization of

North American alternative food networks as reflexive social constructions so that AFSs build shortened supply chains[1] through trust and quality (Goodman 2003).

We begin the chapter with research results from a study of alternative food networks in Toronto, Canada. Our project on the specialty food economy in Greater Toronto was part of a larger body of work that sought to understand regional development opportunities and challenges (Wolfe and Gertler 2001). *Innovation Systems and Economic Development: the role of local and regional clusters in Canada* was a five-year study examining how local networks of firms and supporting infrastructure of institutions, businesses and people interacted to facilitate economic growth. This part of the larger project involved a three-year study (2003 to 2006) that examined how local networks of firms and the supporting infrastructure of institutions, businesses, and people in the Toronto region interacted to facilitate economic growth in the food industry. To get a profile on the industry, we constructed a database of employment and annual sales for 1,465 food firms in the region as well as conducted in-depth interviews with over 100 food firms and other institutions that were thought to be the most innovative.

As one of the most multicultural cities in the world, Toronto is an interesting case study site for food (UNDP 2004). It offers dense material for the study of the alternative food industry and the role that food processors and retailers play as they respond to food fears and define alternative products and networks to address food safety concerns. Our interviews with food processors, retailers, distributors and food experts outside the mainstream food system provide empirical evidence that in part: 1) these food processors and retail firms define themselves in response to consumer food fears about personal health and about more pervasive social concerns; 2) by reacting to these specific concerns, alternative food businesses identify and promote their products in clearly codified terms; and, 3) these codified solutions to food scares provide the mechanisms for the conventional system to co-opt elements of alternative food, and ultimately undermines the integrity of the 'alternative'. Our working definition of 'alternative' began as oppositional in the sense that we took alternative to be anything outside the mainstream, conventional food system. To understand structures in the industrial food processing system, we included interviews with multi-national corporations engaged in food processing in the region (Blay-Palmer and Donald 2006, Donald and Blay-Palmer 2006). These results helped us clarify our understanding of 'alternative' to include organic, ethnic and/or fusion food produced by SMEs. For the purposes of this chapter the results reported have been refined to report only on organic processors, retailers and distributors. Their priority is the delivery of quality food from the local foodshed for local consumption. Our findings lead us to conclude that AFNs need to take stock of what they can offer as part of a larger movement and define themselves in these terms to create strong, independent, resilient, local food systems (Watts et al. 2005).

---

1 Either a physical shortening as in the case of direct sale farmer-eater relations in a local community, or fair trade when a conceptual shortening occurs and consumers develop a deeper appreciation for farmer needs (Goodman 2003).

Fear, Food and Alternative Food Networks

As discussed in earlier chapters, food fears present an interesting intersection between science and society, risk and regulation, externality and corporeality (Kneafsey et al. 2004; Whatmore 2002; Beck 1992). In Chapter 2 we discussed how the fear of contamination and adulteration at the inception of the industrial food system necessitated a central role for government regulations, science and technology. And, in the context of these novel – in scale and pervasiveness – problems, food became increasingly processed. As processing increased and people were more unsure about where their food came from, their uncertainty fed the need to regulate food and supported an increased reliance on the power of even more applied science and technology to guarantee safe food.

Today, food processors diminish the perceived distance between field and fork through packaging, labelling and certification programmes (Watts et al. 2005; Kneafsey et al. 2004). These relationships intend in part to take the fear out of food and offer eaters quality food they can trust (Hinrichs 2000). In some cases, the search for safe food has contributed to an emphasis on food 'quality', trust, shorter food chains and direct buying-selling relationships (Goodman 2003; Hinrichs 2003; Whatmore et al. 2003, Murdoch et al. 2000) – in short the de-industrialization of the food. Alternative food networks however, are rife with contradictions (Maye et al. 2007; Holloway et al. 2007; Jackson et al. 2007; Ilbery and Maye 2005; Guthman 2004a, b). This line of research raises questions about the complexity and hybrid nature of alternative food and has attracted more attention about how to define 'alternative'.

With a view to tackling these definitional challenges, Watts et al. (2005) contrast strong chains – characterized as short, trust-based, personal systems with direct links between producer and consumer – with weak ones. For Watts and his colleagues the difference between the two types of food systems rests on the fact that the weak alternative food chains focus on qualities inherent in the food, while the strong alternative systems define themselves through their networks and connections to a food system. In this sense, the strong connections appear to be more embedded in their food networks. As discussed elsewhere in the book, another important distinction in the literature between different types of food systems is the recognition of complex, nuanced, hybrid food systems with no clear lines between alternative and conventional. Accordingly, food systems are characterized as continual between de-localized/ re-localized (Sonnino and Marsden 2006); strong/weak/hybrid alternative systems/chains of food provision (Watts et al. 2005); local/ global divides (Hinrichs 2003; Winter 2003); or, as being fashioned along production/consumption types to include public, commodity, specialty (quality), and community (local food projects) production/ consumption divides (Ilbery et al. 2005).

However, a missing dimension from these discussions is the role of retailers and processors participating in the 'alternative' food market and the way they mediate and translate social concerns. Although the literature has explored in depth the rural context for producer-consumer links, as well as direct sell relationships, farmers' markets and labelling/ certification schemes (Ilbery and Maye 2005a, b; Watts et al. 2005; Hinrichs 2003) little is known about how food processors and retailers interpret social cues from AFS consumers. The void is especially notable in urban settings. Given the growing importance of urban populations and the clear role that processors and retailers play in mediating food systems, understanding the intersection of these two areas would be useful. To this end, we turn our attention to our interviews with food processors and retailers in Toronto, Canada.

Food Fear in Toronto

As noted in Chapter 1, we originally became interested in trying to understand the role that fear might be playing in the rise of AFNs (especially the increased North American sales of foods labelled organic) when we noticed a pattern of answers in our interviews with various food actors in the Toronto area.[2] In response to questions about the personal and business motivations for starting 'alternative' food companies, key informants told us that fear played a role in shaping their ways of life and businesses.

The responses were an unexpected by-product of our research on the innovative dimensions of food in the North American urban economy. Not all the firms we interviewed were producing organic food. For the purposes of this chapter, we report on those companies and institutions producing organic food as many of these firms were some of the most innovative and dynamic in the region. Most of the people we interviewed were in the business of eating, producing and processing organic food because they wanted to provide what they perceived to be a better food product than what was currently in the mainstream market. Yet what we discovered during the course of our work raised interesting questions about the role of alternative small and medium food enterprises as drivers of innovation (Blay-Palmer and Donald 2006). It also pointed to the way that food facilitates cultural inclusion for new Canadians. It became clear from our research that accessing culturally appropriate food helps people to feel at home in Canada (Donald and Blay-Palmer 2006). In the same way that processors of ethnic food bring the comfort of home to new Canadians by providing culturally appropriate food, organic and other alternative processors, distributors and retailers respond to social needs. In this way, alternative food provides hybrid spaces for processors, distributors and retailers (Watt et al. 2005; Ilbery et al. 2005). Of particular interest for present purposes is how intermediaries in the food chain translate the consumer's need for safety into food products.

The results reported in this chapter are based on 65 interviews. These discussions took place between January 2003 and June 2005 with food producers, distributors, processors and retailers; non-governmental organizations including food security experts and consultants; restaurateurs and chefs; educational institutions; the media; and, municipal, provincial and federal government officials. While the focus of the interviews was on alternative food processing SMEs, 15% of our interviews were with multi-national corporations (MNC) and institutions in the industry. This established the context from the MNC perspective for the larger food processing

---

2 Food anxiety was also an unintended result remarked upon by Kneafsey et al. 2004 in their research on the (re)connection of consumers and producers through 'alternative' food networks.

industry in our study region. The results discussed in this chapter report on a subset of 25 interviews with 5 retailers, 14 processors, 4 distributor/retailers, and 2 producer/ processors. While the majority of interviews were with retailers and intermediaries, two participants were farmers engaged in production, processing and retail activities. Seventeen of the businesses interviewed are certified organic, with the remainder focusing on local food for niche markets. In the rest of this section we present our findings under two sub-headings: defining alternative food processing and retailing; and, codifying the message.

Defining Alternative Food Processing and Retailing

Our analysis revealed that fear emerged as a factor defining the features of the products and businesses of our key informants. In our work, interview subjects identified both the individual consumer's personal fear as well as a more pervasive societal fear as stimulating consumer interest in alternative food.

As Kneafsey et al. (2004) make clear in their assessment of anxiety and alternative food systems, there are two sides to the way these concerns are expressed. First, there are concerns related to 'body and health' – what we have broadly termed personal-based fears, and anxieties about 'locality and community' – what we have called pervasive, social level fears. Retailers and food processors stated that these fears served as focal points for their firms. As one retailer commented,

> We position ourselves against two of the largest bodies of power – the agro-chemical companies and pharmaceutical companies. We are the antithesis of them. People's focus on health food scares, and the need for alternative foods and medicine allows us to offer a choice. (Retailer, April 2003)

Clearly, a key emphasis for this retailer is on alleviating consumer fears. These retailers describe the interaction between personal and societal spaces and fears as a powerful force driving growth in AFNs. Personal and societal forces feed off each other so that consumers turn to AFNs to alleviate their concerns about the corporate-dominated food system. This fear creates gaps for alternative retailers where they can provide a safe and trusted space where agro-chemical and pharmaceutical companies are not. Personal fears also provide the motivation for entrepreneurs to develop alternative food products. A processor of organic jams and condiments offered her personal rationale for avoiding the industrial food system,

> North America is a *giant Petri dish*. [Our] children are sick. How much is being caused by the food we eat? (Organic processor, February 2003, author emphasis added)

The key informant started her business as an alternative in every sense. As a result of personal health issues and an ethical crisis about the financial industry she was working in, she searched out an alternative business opportunity. In this case, her concern for societal well-being, as well as her own health, motivated her to make drastic lifestyle changes. As discussed in Chapter 5, the impacts of outside forces and shocks on the food industry such as Severe Acute Respiratory Syndrome (SARS) and BSE have generated consumer fear about personal health. This fear is compounded by the related perception that science and institutions have failed in their ability to regulate, control and protect the integrity of North American food (Klint-Jensen 2004) or to properly represent and protect the public interest (Concentration of GE corporations: Phillips 2002, Harhoff et al. 2001). The discovery of BSE in Canada in 2003 and in the US in 2005 accelerated the demand for naturally-raised and organic meats. As one organic meat producer, processor and retailer explained, the fall-out from SARS and BSE in Canada has been a boon to his family's organic meat business:

> In the first two weeks of SARS home delivery doubled…Mad Cow disease did wonders for the industry, it was inevitable that consumers were going to start asking more questions, Mad Cow just gave it a prod…In organics we have audit trails, accountability …we can 100% guarantee our product. (Organic Meat producer, processor and retailer, April 2003)

It is the direct link between the eater, farmer and the food that makes the difference. Trust comes from asking the person selling you the meat about where it comes from, how it was raised and killed. This knowledge reassures consumers. The ability to access food from alternative sources allows consumers to cope with the shocks in the mainstream food system and they are able to mitigate their fear about the food they are eating. The fear created by these events push consumers to seek out alternative food provisioning systems.

Faced with the desire to both extricate themselves from the IFS and reconnect with nature, one farm family we interviewed switched from conventional to organic farming and processing:

> We found it [organic farming] was a safe environment to work in. We don't want chemicals for ourselves and our family, for the soil and our present/ future capability to produce food. We were tired of what big industry was telling us to do, to be part of their agenda. We were a tool for them to make money; this was not to our benefit. We want to be independent, and support the natural world to enhance diversity. (Organic producer and processor, March 2003)

The need for balance between social/family, economic and ecological needs is clearly expressed by this farming couple. In addition to changing over to organic production, they also now run an organic processing facility that also serves as a store. Their very popular product is distributed throughout the Toronto area.

The increased demand for trust-based, quality relationships and products is creating the fastest growing sub-sector of the food industry (Minou and Willer 2003). But we also know that agro-food companies from the IFS are filling these spaces as they too profit from the food fears they precipitate (Guthman 2004a, b). Recalling the example of Jolly Time candy apple popcorn, the IFS plays on these fears as a way to carve up the market and create new, niche opportunities (Schlosser 2002). When AFSs translate food fears into codified lists that can be replicated, AFSs provide a directory of factors that the mainstream food can – and does – address.

Codifying Fear

Clearly, there are food retailers and processors who understand and capitalize on the fear that consumers have about their food. The space that these intermediaries occupy in the food chain positions them to act as translators of consumer fears (Callon 1986). From this vantage point, food processors and retailers are able to interpret, focus and consolidate consumer concerns into market opportunities. Preoccupations about food safety are fed back to consumers in codified forms through a myriad of venues including product labelling, advertising, certified standards and websites. As one food distributor describes, the organic food network guarantees that:

> Organically grown foods are produced without the use of synthetic fertilizers, pesticides, herbicides, hormones or antibiotics. Organic certification is the consumer's guarantee that foods are grown and handled according to strict standards that are verified by independent organizations. Certified Organic foods are not irradiated and do not contain genetically modified organisms. (Food distributor, March 2003)

This description of the benefits of organics is a checklist that mainstream food producers can use to fit themselves into this market (Guthman 2004a).

As another facet of the contested nature of alternative food, the 'label' and its credibility have been conditioned through the industrial food system. As we explored through our brief history of food regulation, marketing and retail, the industrial system was founded on the distanced relationship that the label itself, in fact, embodies. In the end, the conventional networks that taught consumers to trust labels and standards bolsters the power of the label on organic food. As also discussed earlier, sophisticated marketing and labelling in concert with government food safety regulations worked to create an image of food safety. And, since the alternative network uses certification standards with attendant labelling to confirm the authenticity of their products, this paves the way for mainstream processors into the commercial space created by fears of their original 'conventional' products, and which the 'alternative' system was – up to that point – otherwise able to exploit by themselves. The existence of standards in both food systems provides a way to scale up for alternative food processors and a way into the AFSs for mainstream food processors. So the challenge becomes, how are AFSs different from the industrial food system and how does this get communicated in a meaningful way to consumers? A retailer in describing their role in the community explained,

> We work tirelessly, advocating fewer and safer pesticides in non-organic foods, in educating our customers about the value of foods produced without harmful or questionable food additives, and we have worked with manufacturers to supply our stores with foods that meet our strict quality standards...We educate our customers about the importance of food safety measures and techniques, including our concerns about irradiation, food borne illnesses, food handling, and material safety. (Retailer, April 2003)

One food processor determined to offer her products as an alternative to the industrialized food and describes them as,

Table 7.1   Fear-based and positive descriptors from key informant advertising and web sites from Toronto and Boston food producers and processors

Firm type	Fear Descriptors	Positive descriptors
Producers	Free of Pesticides and Herbicides	We are committed to providing all you need to live a healthy life
	No synthetic fertilizers, pesticides, herbicides, fungicides or insecticides	Biodynamic, ecological
	Livestock cannot be fed or treated with synthetic antibiotics, growth hormones, genetically modified organisms, colouring agents, animal by-products or medicated feed	Ethical environmental responsibility Community health and well-being
Processors	Wheat-free	Delicious and convenient
	No preservatives	Fresh, flavourful, seasonal
	Cruelty-free personal care products	Local, authentic, ethnic
	Additive free products	We use only the best, all-natural ingredients
	Personal health and safety	Quality
	No chemicals, genetically modified organisms, or artificial fertilizers …no antibiotics or growth promoters are used	We fill the box with the best locally grown seasonal items Healthy, nourishing and full of flavour; health and well-being of the earth
	We do not use any oils, milk, sugar or eggs in any of our breads… all of our breads are made with unbleached and unbromated flour	We carefully select only the finest ingredients. As a result, the bread you love – from our hearth to your home – is hearty, healthy and pure
	No antibiotics or commercial fertilizers…chemical free	Just pure and natural techniques that keep the cows healthy and happy and the dairy products wholesome

...using no chemicals, no additives, no preservatives, no fillers, no food colouring, no food flavouring – all our condiments are GLUTEN FREE, WHEAT FREE, DAIRY FREE, EGG FREE, YEAST FREE, NO SUGAR ADDED, SOY FREE and VEGAN. (Organic processor, February 2003)

The products are defined in terms of what they do not contain. At the same time, however, she is not completely secluded from the mainstream; and in fact one could argue that she is still deeply part of it. Since our interview, her business has entered a new phase, growing to the extent that she is considering outside investors and larger retailers to broaden product exposure and increase economic viability. The emphasis has also shifted from organic to gluten free. The evolution of this business speaks to the conflicted nature of alternative food networks and the up-scaling versus down-scaling tensions that exist as the industry expands and changes. The personal fears that led her into the business are reflected in the societal fears that help her business expand and industrialize. By translating consumer fears into product characteristics, processors and retailers facilitate the co-option process by the industrialized food system.

There is also a complex language and set of messages that emerge as consumer food fears are confronted by food processors. A review and classification of the language used by Toronto and Boston processors and retailers on their web sites exposes a duality when dealing with fear (Tables 7.1). On the one hand, processors and retailers use fear-based language to capture the market and remind consumers about why they need to buy alternative food. On the other hand they offer solutions in terms of human, animal and environmental health. On the negative side, products are described as not containing a range of chemicals, additives and food qualities. On the positive side, the alternative food products and stores offer health and well-being to consumers and the world they live in. By defining the downsides of the industrial system, the alternative system offers consumers a way to recover some of the power they have lost to the IFS. It gives people a way to control what they are putting into their bodies (Whatmore 2002).

But, on a practical level, there is the problem of 'label fatigue' as consumers must constantly read and educate themselves if they are to engage with the food system in a meaningful way (Goodman 2004, 10; Ilbery et al. 2005). Information overload can lead to confusion and the dilution of standards to the lowest possible denominator. The study presented in Chapter 1 by Aubrun and colleagues is worth revisiting. In this study, consumers were found to break their food worlds into 'little pictures' as a way to deal with food scares. And while this allows people to cope, it is less helpful in addressing the complex systemic challenges that were created by the industrial food system. This 'small picture' approach allows MNCs to hive off a piece of the troubling food system and repackage it so consumers feel they are making a difference. McDonald's, for example, now has fair trade coffee in the north eastern United States and Wal-Mart introduced highly processed and packaged organic products, underscoring the contradictory and ambiguous nature of alternative foods and the spaces that labelling open up for AFNs.

While it must be acknowledged that the presence of global players in alternative food networks can expand awareness about the merits of alternative food, the long distances of global industrial food supply chains and the risks these food production systems open up for contamination through large scale farming compromise the essence of alternative food networks as quality, trust-based, shortened supply chains. So there is a need for the alternative food networks to become embedded in alternative food systems (Watts et al. 2005).

Then, if we are to understand the potential for the alternative to become mainstream we need to assess the willingness of consumers to engage in re-localized consumption networks (Holloway and Kneafsey 2000). We also need to determine what local and alternative means for consumers. Caution is needed though so that,

> ...the ways in which such spaces [farmers' markets] become increasingly regulated, and the relationships between these spaces and the spaces of 'conventional' food production and retail, should also be examined in recognition of tendencies towards bureaucratic and capitalist appropriation of what might become alternative economic spaces". (Holloway and Kneafsey 2000: 298; see also DuPuis and Goodman 2005; Donald and Blay-Palmer 2006)

Tacit knowledge, or 'that which is not written down' (Lundvall 2003, 6), is another matter. This is where the industrial food system begins to break down and cannot replicate successes in the alternative food system. Trust and personal connections are built between individuals and they are what consumers are turning to as they try to reconnect with nature, their food and reclaim their power to define the world they live in through food. Marketers know this – it is the rationale for 'Greeters' at Wal-Mart and the personal stories behind archetypical employees. The Wal-Mart website introduces its store and employee feature web sites,

> With more than 1.9 million associates and thousands of stores worldwide, there are countless Wal-Mart stories to be told.

> These videos reflect just a few of the Wal-Mart stories that occur each day – stories that demonstrate our commitment to our customers, each other and the communities we serve. Stories that demonstrate the difference Wal-Mart Stores, Inc. and our associates make for people around the world by helping them save money so they can live better. (Wal-Mart 2007)

By telling individual stories of community outreach during the hurricane Katrina crisis, and telling the stories of the people who work in Wal-Mart, the store uses its people to connect online through video clips with customers so they can,

> Watch how Wal-Mart's Emergency Operations Center provided relief to the gulf shore region following Hurricanes Katrina and Rita.

> Watch how one Wal-Mart store is helping to revitalize a Chicago neighborhood and provide much needed jobs for its residents.

> Watch our dedicated associates give back to the community through charitable organizations in their hometowns. (Wal-Mart 2007)

And in another video segment titled 'Supporting our troops', customers find out that,

> With more than 3,000 Wal-Mart and Sam's Club associates serving the armed forces, see how Wal-Mart is helping to support the troops and their families. (Wal-Mart 2007)

Alternative food processors use the same approach. On their web sites and in their advertising material they also tell their stories. In one example, a Boston organic food delivery company has a photo of the people working for the company (there are twelve at the time the picture was taken). There is also a one-paragraph description for each employee that mentions details about family, children, hobbies and life passions. Wal-Mart is brilliant at what it does. On the one hand, it is the largest, most globalized and vertically integrated retailer in the world, yet on the other hand, through advertising and marketing, the company evokes a real home-town legitimacy that has the affect of making it a 'trusted' main street-like company for some people. However, these images disintegrate as stores are closed when workers try to bring unions into the company and main streets and local businesses die in the wake of 'Wal-Martization'. So while the 1.9 million Wal-Mart associates try to connect with their customers in a personal way, the small alternative food processors and producers actually do connect and build lasting, genuine, trust-based personal relationships with the people who eat their food. This leads us to the point of the book – that having identified fear as a defining factor of alternative food networks, it is possible to circumnavigate fear to embrace hope. Since the conventional food system has created problems that are increasingly difficult to contain and regulate, there is a need to consider other ways of doing food. We suggest by clarifying the non-codified strengths of the alternative system, we can reinforce its success. The alternative system clearly faces great challenges in singling itself and the advantages that it offers out from the conventional system in the face relentless efforts by the latter to co-opt every market niche that the alternative system identifies. In the next chapter we offer some hopeful examples and a template for food systems that builds upon cooperation, not competition, as the starting point for sustainable food systems.

Chapter 8

Creating Mutual Food Systems

Until recently I was ignorant about what organic meant. I thought it was 'nothing bad', so for me it was food not ridden with GMOs, antibiotics, herbicides, pesticides, chemical fertilizers –I thought of these as bad. What I didn't understand was that organic has to do with positive things not just bad things. ...I learned to realize, to understand that organic isn't just about NOT spraying, but it is about the benefits. When I go to the farmers' market, I used to ask, "Do you spray?" Now I ask farmers "What do you do to enrich the land?" They tell me "We use manure and seaweed". I realized that organic means giving back to the land, not just avoiding toxics. (Wisconsin/ PEI resident, September 2007)

Free market ideologues, evolutionary economists, and other experts assume we are evolving in the direction of a more equitable world. The theory, founded in neo-classical economic thinking, is that we will achieve this ideal through individual choices that collectively signal the market about the best direction for the greater good. As the case of food demonstrates, this market-based approach does not necessarily provide more good. In fact, from the perspective of malnourished citizens, impoverished farmers and the stressed environment, it creates increasingly dysfunctional relationships and has accentuated the disparities between winners/losers, rich/poor, and north/south. Given these shortcomings, it is prudent to consider whether this competitive model is indeed valid for the 'intimate commodity' or whether other models would be more appropriate. Consistent with the questions being explored in biology raised in Chapter 1, it may be timely to reflect on the benefits of cooperative social structures as the basis for healthy food systems (Clausheide and Courtenay-Hall 2007; Harvey 2000; McLeod 1976). As the quote introducing this chapter suggests, perhaps there is value in appreciating the benefits of doing things from a positive and hopeful perspective.

This book has presented a number of challenges that have emerged in the IFS. By looking at food systems through the lens of fear, I have tried to clarify the places where the food system is broken as a way to find a path away from industrial food systems towards a new way of relating to food. In the first part of the book, I examined the structures that underpin our food system. The use of public research dollars, policy and regulation in North America to privilege large-scale production over small, and private over public interests was evident as a problem early in the book. The way the R&D and policy environment is structured necessitates increasingly complex regulatory systems such as HACCP. In the rush to ensure global trade flows, the resulting bureaucracy makes the food system more costly and erects trade barriers for small producers and processors. This has the effect of scaling-up the food system and making it more opaque. On the marketing and retail side, much of the emphasis is on the creation of market share for highly processed food products. The discussion of food retail made the point that a strong effort is expended to build trust – this is after all the point of branding a product and of building brand loyalty as early as possible. Accordingly, public perception is a key contributor to product value in the industry. This is especially true in the case of highly processed food, where manufactured convenience and taste (largely through sugar and salt) are where the industry makes the most profit. Food marketers and retailers also lay the groundwork for complex and ambivalent relationships between people and their food as corporations exploit the 'fear factor' as a tool to raise their profile. The existence of so much ambivalence and complexity in the IFS necessarily increases fear and uncertainty. The under-represented place of local ecologies in food systems and the difficulty farmers have in trying to balance so many conflicting priorities was explored in Chapter 4. With the reality of so few farmers left in Canada and the US, we are left with a growing sense of apprehension about our ability to develop the supply side of a new food system. Disease and chemical contamination exposed another facet of complexity in the food system as networks of food are mobilized to preserve networks of capital at the expense of human and animal health. Through the lens of Alar, salmonella, Mad Cow and avian flu the question of scale emerged from the perspective of disease and food. It became obvious that the bigger the system, the more complex and riskier it became from the point of view of disease and contamination.

In this climate of compromised food relationships, we explored the attraction of eating local organic as a space where consumers can control an aspect of their lives. By deciding on what they are going to eat, consumers bypass the political-economy of industrial food systems and opt to consume from a food culture where they have assurances of safety and quality. These assurances sometimes come from confidence in certified product, or more deeply from trust-based, face-to-face interactions with local food producers, distributors and processors. In this way they bypass the threats inherent to the industrial food system and the broader society at large in the wake of 9-11. Fear also emerged as a defining factor for food processors. Once again, a critical ingredient for consumers in constructing their food system is trust.

Essentially then our analysis of the IFS and some aspects of AFS points to an overly complex system that has grown so large it is becoming unmanageable as it creates problems it can no longer resolve. These insights raise questions about the role of policy as an avenue to support the AFS which identifies and addresses real needs and rapidly increasing demand, rather than supporting the IFS which compounds the problems.

It is important to pause and reiterate an important assumption. Although we have spoken about industrial and alternative food systems at some points in the book as if they are somehow mutually exclusive this is often not – as laid out in Chapter 1 – the case in reality. While it is possible to find cases of for example, only organic or only conventional food production/consumption linkages, there are usually no clear boundaries between the two systems. More often it is the case that the two systems overlap. At the very least, they are both contained within the same regulatory frameworks that serve to reinforce and constrain certain features of both systems. Beyond recognizing that the regulatory system does not support the AFS in addition to the dilution of its efforts in managing the mainstream system, there are other inherent issues. It is equally necessary to recognize the contradictions and inconsistencies that exist within AFSs. As discussed in earlier chapters, these 'alternative' systems represent ambivalent and contested spaces in the food production landscape as aspects of production, distribution, processing, retail and consumption may be subsumed into the existing industrial hegemony (Maye et al. 2007). We confirmed that the way intermediaries in the food system 'envisage' (Murdoch 2006) food for consumers facilitates this process. By capturing and then codifying consumer concerns, producers, regulators, processors, marketers and retailers translate food fear into food product. This can add to growing inequalities as elite consumers for example eat "'designer' organic vegetables that get shipped around the world in a sophisticated 'cool chain'" (Watts and Goodman 1997, 3). As mentioned earlier, Goodman (2004) and others (Maye et al. 2007; Morgan et al. 2006; Donald and Blay-Palmer 2006; Hinrichs 2000, 2003; Winter 2003) advocate on behalf of the missing guests at the table as they speak for equality and inclusion within alternative food spaces. Guthman (2007, 2004a, b) is very vocal about the cooption of the alternative food system by agri-business as the dominant food industry reduces organics to 'yuppie chow'.

However, as argued in Chapter 1, there is merit in recognizing that separate food systems exist as this provides us with the ability to analyze their strengths, weaknesses and effects. On the one hand, we have the industrial food system that is constructed on production principles similar to those used to produce any other commodity. The tendency is to apply technology and science to resolve problems, while the economic strategy leans towards a neo-liberal globalization approach. On the other hand, is the emerging system of food production that privileges short supply chains, trust, and ecological production principles. Production-consumption links are as direct as possible, and trust is a key ingredient in consumer confidence in their food.

Within the complex set of relationships that include personal and social constructs, food retail, marketing and regulation, science and safety, and the industrialization/ de-industrialization of food, this book points to fear as one manifestation of the ambivalent relationship people have with their food. Fear exposes spaces in the food system where trust has been violated in one way or another. Preceding chapters have explored the creation of regulations and oversight processes for the increasingly complex industrial food system as necessary to protect consumers from everything from food adulteration and disease, to degraded local ecologies and communities. We have also seen that the complexity of the system means increasingly that gaps emerge that are not or cannot be regulated. It is no wonder then that people fear their food. However, while acknowledging these imperfections, the research presented in this book and that of others demonstrates that hybrid food systems can reduce uncertainty and address other consumer concerns if they provide consumers with food they can trust (Maye et al. 2007; Hinrichs 2000; Murdoch et al. 2000).

On a more positive note, our research confirms the pivotal role of trust between consumers and other actors in the food system as a platform on which to build new food relationships. Thinking about trust-based, cooperative communities raises the potential of *mutual*[1] relations of food where a 'group of people would cooperate towards a common goal or common good' (Oxford English Dictionary online). Food fears have served as a useful starting point for the AFS as they have provoked consumers to search out 'safe' food. The quest links people in search of the same reassurance (similar to Harvey's Utopian vision, 2000) and leads to "alternative visions" (Kneafsey et al. 2004, 8). If food consumption goals are to go beyond 'fuelling' the body, to include elements of trust, community, and social responsibility, then there must be a stronger element of mutuality in food production, distribution, sale, and consumption.

The first part of this book suggests that a cooperative community-based food system may indeed be worthy of consideration. But one question that arises from this line of thinking is where would this change come from? Part of the answer can be found in Suzanne Friedberg's insightful book *French Beans and Food Scares* (2004) where she suggests that windows of opportunity have been pried open through which civil society and NGOs are exacting accountability from global capital so that,

> ...intense market pressures and the mutual need for a good 'brand profile' have, somewhat paradoxically, helped certain non-profit advocacy groups acquire a measure of power [and]....offer insights into the potential of civil-society movements that demand accountability from all kinds of globalized brand-name capital. (Friedberg 2004, 514)

To elaborate how to move the alternative system beyond its current status, Patricia Allen (2004), building on the work of Raymond Williams (1973) offers the ideas of 'alternative' and 'oppositional' as useful guiding ideas in trying to answer this question. Alternative activities include people or groups operating outside the mainstream. Alternative initiatives tend to be small scale and issue specific – in the cases we have looked at in this book, fear-based. They also tend to operate at a more superficial level when addressing an issue, provide a less critical perspective and are shorter term in focus. As well, they may not address environmental and social equity issues. By contrast, oppositional approaches try to change the structure of the system. They tend to be large scale and envision radical change with long-term implications and thinking. Oppositional approaches address the danger that incremental reformism – that tends to typify the alternative approach – does not really change the fundamentals and can result in accommodation as opposed to meaningful change. Oppositional approaches on the other hand can lead to transformational change.

The new relationships being structured throughout the food chain from field to fork between actors such as NGOs and civil society movements challenge aspects of the IFS (Maye et al. 2007; Morgan et al. 2006; Sumner 2005; Allen 2004). And the reasons behind the increased profile of these systems provoke us to consider the validity and robustness of our existing system. Halkier (2004) offers a succinct summary of factors that are relevant for understanding the politics of food and the

---

1 Mutual is used here in the sense of mutualism from biology. Mutualism is defined as, "The action or practice of a group of people in cooperating towards a common goal and for the common good" (Oxford English Dictionary online).

ability of people to effect change. She puts forward three themes from political sociology as the basis for her analysis. The first is *agency* which she defines as "the capacity of citizens to act"; the second is *community* as "the degree to which citizens experience a sense of belonging to a collectivity that is related to their actions"; and, the third theme is the ability to *influence* as "the ways in which citizens are capable of making a difference to societal problems by their actions" (Halkier 2004, 23-24). In the next sections we explore three case studies that provide insights into the challenges and opportunities created by the IFS. These examples allow us to understand the potential for agency, community and influence in the context of hybrid solutions that combine aspects of alternative and oppositional approaches to building community food systems (Allen 2004).

Innovative approaches to institutional and regulatory structures by members of NGOs and civil society drive these changes (Friedberg 2004). But our further contemplation of the whole matter should be undertaken in the context of the question "Is it time to take policy initiatives to foster the evolution of the AFS out of its "alternative' status?" If much of what drives the alternative food system development is reaction to the 'conventional' IFS, and if the AFS is growing faster than the IFS, why does policy support the perpetuation of the IFS? Instead of having policy that helps the IFS to sustain its model while slapping band-aid product innovations on the problems it creates, should policy instead support a more fundamental shift in the food system, and underwrite what people want and need rather than what the existing system has to offer?

The next section begins with the dairy industry as an example of farmers and consumers operating outside existing regulatory systems to forge their own food and new community linkages. Second we explore the overwhelming success of school meal programs as examples of new public procurement strategies. These new ways of 'doing food' build opportunities to stabilize demand for farmers and get healthier food to our children. The last case studies focus our attention on community minded alternative food programs that help people who are socially marginalized get access to healthier food and build a sense of community through cooperative initiatives. In the final part of this chapter we distill the common elements from these case studies to identify threads that could provide the foundation for a more hopeful food system by capitalizing on elements of mutuality.

Institutional Turns

One way to overcome fear is to go local, scale down, and establish reflexive food relationships that emphasize quality and freshness. That is, to eat from our 'foodshed' (Kloppenburg et al. 1996). Shorter food chains and local producer-consumer connections signal an important shift in food systems as consumer driven quality offers the opportunity to re-empower the eaters. Policy initiatives at the macro level, such as the EU Public Sector Directive 26 (Sassatelli 2006), and civic initiatives at smaller scales, such as student driven university buy-local programs in the US and Canada (for example, Yale 2006) provide templates to re-design the way people connect with their food. These initiatives offer the chance to localize power through food and present practical solutions for local economic development challenges (for example, James Kirwan and Carolyn Foster's study of the Cornwall initiative, 2007).

Product-place-process strategies described by Ilbery et al. (2005) offer guidelines to members of alternative food communities as they seek to negotiate alternative food arrangements as a means to move beyond consumer anxieties and create the critical mass needed to replace the IFS. However, addressing consumer quality issues as part of the move towards hope-based food systems is a complex task. Enticott's (2003) work on un-pasteurized milk (UPM) in the UK is useful in this context as it points to the way that tensions between dystopic and utopic consumer visions of food systems are resolved as consumers place their faith in locally emebedded 'common knowledge', and not 'scientific' approaches such as pasteurization. Other research shows that consumers who buy food boxes balance out seasonal variability with food "production practices, freshness and origin" (Lamine 2005, 341). In these cases, consumers decide on levels of quality and safety they can tolerate, find a balance that meets their set of priorities and seek out the best food alternatives to meet their needs. This process continues to evolve as it is reflexive and iterative. It also requires the ability to work with(in) the regulatory system in a creative manner and have trust in the personally constructed food system.

Many innovative producers and consumers meet their needs within the food system by operating below the regulatory radar to create new food relationships. The Canadian dairy industry provides interesting insights into the benefits of innovative approaches to food connections and the creation of new food relations. As a food product, fresh milk has a limited shelf life and so tends to be produced within a city's peri-urban and rural fringes. Since a herd of cows can be milked throughout the year, they are not subject to the seasonal variations of produce, crops and other fresh foodstuffs. As a result dairy offers a unique opportunity to build consistent, local production-consumption links. It also provides a window into both the industrial and alternative food systems as there are robust conventional and alternative dairy networks.

A recent example from the industrial milk system points out how producing food on a large scale can create problems and questionable practices. In Quebec, a provincial Ministry of Agriculture report presented findings about collection processes of bulk-transported milk that allow cow hair, bugs, traces of feces and hay into the milk. The response by the provincial milk association's vice president, the Conseil de l'industrie latiere du Quebec, was, "We filter the milk. We pasteurize it. We homogenize it so that, at the end, the product is completely safe for the consumer". (CBC 2007a) According to this spokesperson for the industrial milk system, safe food lies in our ability to take bulk quantities of food from many sources and sanitize them. The goal is not to prevent contamination – in this case bugs, feces and hair in milk – but to deal with the effects from the extended complex production, processing and distribution system.

By way of contrast, we present three case studies to help us understand the dairy sector from the perspective of people operating outside or in opposition to the industrial food production system as they seek different ways of providing milk products. Each example tells a very personal story of individuals confronting the conventional food system and how they reconcile its challenges in their own ways. These examples are provided to illustrate the challenges that people face at the micro level on a daily basis. The first, in Wisconsin, is an example of innovative consumer-farmer relations; the second near Toronto, Canada tells the story of a farmer who is fighting existing regulations in order to make raw milk available to consumers; and, in the third case we meet farmers and cheese makers who use sheep milk to produce artisan cheeses. These examples were selected as each one brings to light important features for a new food system. The case studies begin with Wisconsin.

> Wisconsin has the second highest number of organic farms in the US, with 659 certified organic farms on over 91,000 acres of cultivated land and an additional 28,000 acres in pasture. Remarkably, Wisconsin produces one third of the US organic milk on organic farms that generate average revenues of $150,000. (Miller et al. 2006)

But to do things outside the established system – even the established organic system – requires ingenuity. It is also necessary to escape the attention of regulators. This is the case for a group of raw milk drinkers from Wisconsin.

> The way it works now is that if you own a cow you can do what you please. So, several families share a cow. At the moment this way you can buy un-pasteurized, un-homogenized milk. If you try to make a profit then you need to comply with the regulations. Every year there is a threat to what we can get access to…I want to drink this because of the benefits of whole, raw milk. What people are doing is denaturing the milk, and removing the nutrients and replacing them with synthetic nutrients that cannot be as good as milk in its natural state. (Raw milk consumer, September 2007)

This key informant drew parallels between the anti-smoking campaigns and the shift to alternative foods. Having been involved in the anti-smoking campaigns in Wisconsin, she was able to identify both similarities and differences between smoking and the fast food we eat today. She points to lessons learned and future challenges,

> Every generation seems more sickly than the generation before…We are what we eat. People are getting fat as they are malnourished and their bodies are hungry for vitamins and nutritious food. In the 1970s when the anti-smoking initiative got started it seemed to be so hopeless at the time. I can remember getting kicked out of a bookstore for suggesting the owner should ban smoking. I have seen a complete reversal over time. The world has turned completely around, so then why should it not do the same on organic food, why wouldn't there be a shift in eating. There has been for the kind of oils we eat, there is an awareness that is growing. But organics is going to be harder as it involves economics, spending more money. The average person will have to spend more money to eat healthier food – that is the biggest impediment to this movement that wasn't part of the anti-smoking movement. Until we can drive down the price of organic food we still have a problem. (Raw milk consumer, September 2007)

This raw milk drinker describes unexpected community benefits that emerged recently as Wisconsin farmers faced extreme flooding levels. As in many North American farm communities this year, the farmers were hit first with drought that baked their soil and then with heavy rains. In one August weekend parts of Wisconsin received over one foot of precipitation. Flash floods and heavy rain caused three deaths and an estimated $48 million in damages was sustained in twelve Wisconsin counties. The governor declared a state of emergency in five counties so farmers could qualify for federal aid (FEMA 2007). And the problems did not abate as flood warnings continued into the middle of September. However, the reaction to the flooding points to a deep sense of community as people sought out ways to help *their* farmers. Speaking about the floods one key informant reported,

> There were benefits [events] to raise money for farmers, and a commitment for some federal funds. We can see people developing relationships with the land because they are developing relationships with *their* farmers. In our case, we look forward to seeing our favourite farmers every week. The farmers' markets allow us to develop relationships with our farmers, to develop a concern for their welfare. When we asked how we could help, our farmers told us, "We are having a dinner, you can help by supporting the dinner". So we bought tickets for $65 each. The floods have created more appreciation for local farmers and the land. People are more connected. (Wisconsin resident, September 2007)

Another excellent example is a recent 'Tour de CSA' in the Madison area. As promotional material claimed, cyclists who participated had the chance to,

> Tour Dane County's beautiful farm country and visit three Community Supported Agriculture farms. Your ride will be fueled with gourmet local food, including breakfast, snacks, and a three-course lunch prepared by Underground Food Collective, Café Soleil and other star cooks from our local food system – all included in the $35 registration fee.
>
> All proceeds, including pledges collected by bike riders, will support Madison Area Community Supported Agriculture Coalition's Partner Shares Program. Since Partner Shares' inception in 1997, over 2,300 underserved and low-income households have received fresh local food. (Tour de CSA 2007)

The bike tour attracted 350 participants. The event was sold out two weeks in advance and many eager participants had to be turned away. The goal of the tour was to have consumers meet farmers and learn more about local food while the proceeds went to support low-income families who are included as part of a healthy, local food system. Building from its roots in the community that go back over ten years, it was a trust-building, educational fundraiser that linked eaters and farmers.

In Canada, near Toronto a similar outpouring of support occurred when farmer Michael Schmidt had his family farm and home searched, and milking equipment confiscated. The raid was carried out by 20 armed Ontario Ministry of Natural Resources officers. Michael Schmidt has been charged with operating an illegal dairy. His first hearing was set for October 2007. Michael Schmidt had been allowing families who own a share in a cow to have access to raw milk. A share costs $300. Thereafter, the owners are able to buy milk (what Schmidt calls Rich And Wholesome (RAW) milk) for $2.00 per liter. The problem lies in provincial and federal laws from the 1930s that prohibit the sale or giving away of unpasteurized milk (Perkel 2007). Schmidt is quoted in a CBC interview explaining the reason for embarking on a month long hunger strike to protest the government's position on raw milk,

This is a battle out of principle. This is a battle that people gain respect again for the farmer…When there is a law which is unjust and which claims that the milk is OK as long as the farmer drinks it, but the milk is dangerous as soon as it crosses the road, that law doesn't make sense.

Local chefs and cow shareholders rallied behind Schmidt as he tackled what he sees as injustice. One of Schmidt's shareholders stated, "We prefer to buy our foods through farmers; we want to have a relationship with the farmer. This is a foolproof system: to buy food from people you know and trust" (McGill in CBC 2007b).

Another challenge to Canadian dairy farmers is the supply-managed milk production system. While farmers who own quota under the program are guaranteed a fair price for their milk, farmers are obligated to buy quota. Farmers are able to do this during a monthly exchange period. For the September 2007 Dairy Farmers of Ontario quota exchange, the clearing price for quota per cow was $29,200 (DFO 2007).[2] In the previous eight months the clearing price fluctuated between $26,837 (June 2007) and $30,501 (March 2007). Actual prices paid for quota during an exchange period are within $150 to $650 per animal of the clearing price. Although the quota system protects farmers in the system, it acts as a barrier to entry for many. As a result of the high costs of buying quota for milk cows, some farmers are turning to goats and sheep as a milk source. These animals are outside the quota system and their milk can be used to make high quality artisan cheeses. Several farmers and cheese makers are involved in these new collaborative efforts.

On Prince Edward Island in eastern Canada, the Veinot family's sheep produce milk for cheese. The cheese itself is made off-island and brought back by the Veinot's to the weekly farmers' market in Charlottetown. Margaret, Alistair and their four children have been searching for a balanced way to farm for the last twenty years. Margaret, Alistair and their first child who was then one year old, moved to their farm in 1987. They had 50 turkeys, a horse and a cow and had convinced a leasing company to buy what would eventually become their farm. They were then able to rent the property, work the land and live in the house. They had no machinery, and the old house and barn needed attention.

Having watched her parents struggle on her childhood farm as she grew up, Margaret knew that she did, "not want to get into the cycle of paying chemical and fertilizer bills, and we wanted to treat the land properly". So they decided to farm organically. Over the years, the Veinot's have raised and milked Jersey dairy cows as well as managing relatively larger or smaller vegetable plots and have sold the produce directly from their farm as well as at the farmers' market. They each held down at least one job off the farm to make ends meet – still it was ten years before they could qualify for a mortgage and start to buy the land they had lived on as tenants for ten years.

As the first farm certified organic through OCIA on the Island they had a lot of ground to break. They also had to figure out how to farm organically. As well, Alistair volunteered for ten years on many boards to help people understand the

---

2 The quota is actually sold per kilogram of milk per day. This is the average production for a dairy cow in Ontario.

relevance of organic farming. But as Margaret explained, "…you do things to the detriment of your own operation". The milk operation was not bringing in a reliable income stream, so they decided to sell their quota and gradually sell the animals for meat. "Then BSE hit in 2003 and there were major changes with the border shut, so we had no market for our Jersey cows. The Canadian market was glutted, the market collapsed so we had animals that were milk fed for veal and that market collapsed. Before BSE we were buying calves for $150 fed them milk and sold them for $300. After BSE the calves could only be sold for $100, so we decided to go into sheep".

Today, they milk their own sheep. The milk from their ewes is processed off-farm into cheese that they sell to dedicated customers at the Charlottetown farmers' market,

> There was a demand for the sheep cheese so then we started to sell other artisan and specialty cheeses. We deal with more unusual type of cheese – the kinds and types we carry depend on what customers want, we find people really want local, made in PEI first, then Maritimes, then Canada, but the general population wants to buy very local.

Margaret also noted that,

> From the tourism aspect, people want something they can't get at home, they want something unique to PEI. We have a local clientele who want to support local people, they want to support the people making the product, not big business. We found the same thing in vegetables. People want to buy something from the farm they know. We have always had an open door policy – we do regular tours on the farm, we just did a tour for about 60 people. People are getting more interested in food again, might be part of Chinese food scares, people want to know about where their food is coming from.

When asked to comment on their relationships with their customers, Margaret remarked,

> I think it gets to be a personal connection. We have always had a fairly open door policy. You know, if you want to see how it's done, come out to the farm, we haven't been off limits at all. And then we talk to people, we become interested in each other's lives, you develop a relationship and that maybe builds trust. If you are interested in their lives and listening and talking, you get to know your customers and you notice if they aren't there and you ask about where they were, if they were ill or traveling. We develop a personal relationship with people.

In addition to building strong ties to their community the Veinots have also been innovators in environmental stewardship. Margaret describes Alistair and herself as "both conscious about where we live, we don't use chemicals pesticides or fertilizers". In their efforts to manage the manure and waterways on their land, first they fenced their animals out of the ponds and river. Along with the not-for-profit organization Ducks Unlimited, they experimented with building a series of ponds to filter the water. This technique has become a standard practice now for water filtration as it allows each pond to remove more and more nutrients as the water gradually makes its way into the river on their land.

The final case study is from southwestern Ontario where a severe health condition led one woman to find food she could eat without becoming ill and eventually to sell artisan cheese at the Guelph Farmers' Market.[3] Pauline Creedy was hospitalized due to the damaging effects of over 400 food allergies on her body. After five years and a focused effort to rebuild her own health, Pauline had identified some facets of the food system that had been making her ill. As it turned out, sheep and goat's milk offered a large part of the solution to her health challenges. Her search for sheep cheese led her to Montfort Dairy and Ruth Klahsen. This dynamic veteran gourmet chef and teacher, is the maker of artisanal cheeses. Ruth's motivation for using sheep's milk can be explained as,

...sheep's milk is a wonderful primary ingredient for the production of extraordinary artisanal cheeses...and that's what we make. We make cheese for people addicted to the simplicity of a well-made food, food that hasn't been mucked with or industrialized into a thing on a shelf, an anonymous product more convenient than nourishing and flavourful. (Monforte 2007)

Her goal was to do things differently as, "Industrial cheese is made in vast batches, all too often with many of the distinguishing traits, tastes, and, yes, the sensuality, machined right out". Ruth emphasizes the importance of food as a social celebration,

We're proponents of slow food, of taking time to eat, in concert with nature and in the company of friends and family, of reviving a sociability in food that many think we're close to losing. And we believe the most beautiful is almost always the most simple. (Monforte 2007)

Their emphasis is on producing a quality product,

Monforte's unique approach to cheesemaking blends 21st century European cheesemaking techniques and centuries-old craft traditions with local Mennonite-produced sheep's milk, yielding cheeses of superb smoothness, well-rounded, almost sweet taste, and in styles both classic and innovative. (Monforte 2007)

Monforte is committed to producing cheese in an ethical, sustainable and respectful way in harmony with the local community by,

- producing memorable all-natural sheep cheeses of outstanding quality
- trading fairly with our Mennonite shepherds
- sharing the ideals of the international Slow Food movement
- respecting our craft and continually striving to refine our approach to cheese-making
- building a viable sheep dairy industry viable for all participants: shepherds, cheesemakers, and consumers alike
- respecting and ensuring the welfare of the sheep whose milk we use. (Monforte 2007)

3 More information is available at: http://guelphfarmersmarket.com/.

However, there are myriad production and selling regulations, and the mundane, quotidian concerns at the micro level threaten to slow or even squash the initiative. For example, to spread the word about the benefits of her cheeses, Pauline offers samples to market-goers. The high cost of the cheese makes this an expensive undertaking. Questions that preoccupy Pauline include the challenge of increasing the availability of the cheese, while still being able to supply the same quality product. She identified the need for more infrastructure to support the industry as it grows. In the end though, Pauline reported that the sheep cheese, "make my booth work, the cheese is so popular that the business has grown". She attributes her success to the interaction between her customers and the high quality cheese,

> I find that the customers love the variety, they love the ideas that the cheese is made with loving care, the way they would like to see cheese made more often. We have such wonderful customers. They are caring, concerned about the community, the kind of people you are glad to have as neighbours. (Creedy 2007)

Again, we see the importance of trust, quality and community. The passion for food that Ruth, Pauline and others express for food spills into the community and closes the loop on producing and eating food with care.

Exercising the Power of Schools

School meal programmes are an emerging feature of the new food landscape and offer multiple solutions to the dilemmas created by the industrial food system. A few case studies help to illustrate the benefits that emerge from these new institutional relationships. Thinking from the micro to the macro, these programs can be divided into three categories: school gardens, school meals and institutional buying. Landmark cases such as Alice Waters' *Edible Schoolyard* project broke ground and provide inspiration for other food activists. Begun in 1995, the *Edible Schoolyard* provides food growing and cooking opportunities to middle school students at Martin Luther King Junior Middle School in Berkeley California. The garden takes up what was once an acre of asphalt-covered parking lot. Cooking is done on-site in a refurbished kitchen. Children learn to prepare food, and learn to eat as a social experience. Class material for the grade six to eight students includes *Principles of Ecology* that teaches students about where their food comes from and to respect the land and its living systems. Students learn about closed loop ecological systems as they use compost from previous years to nourish beds before planting the seeds and tending to their gardens. Students also learn about natural cycles and the seasonality of food (The Edible Schoolyard 2006a). The school describes its students as,

> King's student population averages 850-900 students each year. Of those, 40% qualify for the federal Free or Reduced Lunch Program. Students are 36% African American, 34% Caucasian, 20% Latino, and 9% Asian American. Twenty-two languages are spoken here (The Edible Schoolyard 2006b).

The teachers continue to be engaged with the Schoolyard as a teaching opportunity. Over 80% of teachers use the garden and the kitchen in their curriculum. The current principal, Kit Pappenheimer describes the importance of the Edible Schoolyard to her school community as a chance for students to experience,

> Cultivating the earth, watching chickens lay eggs, gathering their harvest, and cooking their own food; these lessons literally bring school alive for our students. I believe connecting the food they eat with the planet they live on is essential to a child's understanding of the world today. The Edible Schoolyard is a hands-on, creative, and equitable opportunity for all our students to grow! (Edible Schoolyard 2006b)

The UK has embraced the School Food movement and has been providing leadership in this area both through its successes and lessons learned. With celebrity chef Jamie Oliver heading up the charge, and an army of 9,000 chefs in his wake, schools across the UK taught students how to cook during the British Food Fortnight in September to October 2007. While the 'dinner ladies' who prepare school meals have been vocal about the need for more help to prepare scratch meals, and with not all children embracing the changes, UK schools are tackling the problem of empty calories and tasteless food (Morgan et al. 2006). The UK government is sensitive to the challenges and has pledged £280 million for kitchen and training upgrades to facilitate the changes that healthy meals involve. The most recent innovations in UK schools come from Heart of England Fine Foods (HEFF) and involve upgrading vending machines to serve healthy food and drinks including water, juice, milk, fruit juices, fruit, yogurt, and nuts. Innovations would also include juice bars that would offer smoothies made from local fruit.

Indeed, school meals are shaping up to be one cornerstone of a new food movement, "in placeless foodscapes such as the UK and the US, creative public procurement could be the most important single factor in fashioning food localization" (Morgan et al. 2006, 196; Morgan and Morley 2001, 2003). With over 3.25 million school meals served daily in the UK there is a "critical opportunity to place the procurement of school food – a fifth of the £1.8 billion total spent by the English public sector on food and catering – on a truly sustainable footing" (Pearce et al. 2005). In continental Europe, school meal programs are offered in many countries including the Czech Republic, Estonia, Finland, Hungary, Italy, the Netherlands, Spain and Sweden. Italy is a leader in this area. In 2001, School Food Director Silvana Sari, the Mayor of Rome, and Counselor of Education implemented a city-wide school meal program. This program feeds over 140,000 meals cooked from scratch every day using primarily fresh, local and when possible organic ingredients. While price is a criterion in selecting food service providers,

> ...food quality and food service infrastructure are also important criteria. Food quality considerations include place of origin, food miles traveled, organic production, fair trade, and products from specially-designated regions (e.g., Parmesan cheese must be exclusively from the parmigiano reggiano region). Infrastructure improvements include kitchen and dining room upgrades, training and education for staff and teachers, and a well-organized and fully qualified food service staff. (Liquori 2007, 1)

As Liquori points out, the more than $7 billion spent in the US every year on school meals could be better leveraged to the benefit of local communities. Strides are being made in the US as well. In 2005–2006 over 30 million children were fed through the National School Lunch Program. The challenge in the States is not the quantity, it is the quality of food. Schools were reimbursed only $1.35 for breakfasts and $2.47 for lunches in 2005, however, efforts are being made to improve the quality of food served in schools. According to a 2006 National Farm to School Program survey, there are over 1,000 school districts in more than 35 states with farm to school programmes. For example, in New York state over $1.5 million of fresh fruits and vegetables were purchased, with 60% of the produce locally sourced. Interestingly, in 1996 the purchasing of fresh fruit and vegetables for schools began to be consolidated through the Department of Defense (DoD). Under the 2002 Farm Bill $50 million was made available to schools through their commodity funds allocations so they can purchase fresh fruits and vegetables from the US DoD (USDA 2007).

In Canada, several versions of school meal programs are underway. While there is no one studying the rapid growth of school meal programs, the numbers are definitely increasing. A recent informal survey of school meal programmes uncovered an impressive level of activity in Canada where there are breakfast clubs and fresh fruit and vegetable programs in primary, and some secondary schools in British Columbia, Nova Scotia, Nunavut, North West Territories, Ontario, and Quebec. Program goals range from linking schools up with local farmers to nutrition education. Each program is unique and reflects the individual needs of the community and the strengths of the people involved. These grassroots initiatives are examples of civic driven institutional reform in action (Sumner 2005). One program, in Thunder Bay Ontario, provides healthy food to students and uses local food as much as possible. Given the northern location of the community, this can be challenging. The coordinator Jordan Kennie, a former primary school teacher, brings school meal experience from teaching and parenting and a passion for wholesome, tasty food to her role in the Breakfast Program. As Jordan explains, "...once I got involved in Slow Food, I thought it would be a neat idea to focus on local and homemade and give kids access to a wider variety of tastes and foods". Jordan does all of the sourcing and most of the baking herself. She gets food from the Red Cross and a buys food from a local grocery store but the emphasis is on using local food as much as possible. So Jordan uses local honey as a sweetener, whole grains, local berries and meats. Although she uses some unexpected ingredients (e.g. tofu in muffins) the kids know the food tastes great, and she has earned their trust so they will try new food and encourage their friends to do the same. She plans for the meals year round. In the summer, she is picking and freezing berries for muffins and pancakes in the cold months so the kids can have fresh tasting, quality food throughout the school year. In the fall, a local, organic farmer donates pumpkins to the food bank. After being roasted in a stone oven owned by a local baker, they are made into 90 pies for Christmas food baskets. Any leftover roasted pumpkin is frozen for School Breakfast muffins. There are Slow Food baking sessions, and people Jordan trusts will donate a dozen or so muffins to the program on an ad hoc basis. Jordan makes as much as possible from scratch, including granola so she knows the children are getting balanced healthy food. She avoids processed food and opts for nutritious food, not just filler.

The school that houses the Breakfast Program draws its students from across the city and the surrounding rural communities. The school serves a broad demographic. The Breakfast Program is filling an important gap. Jordan reported one conversation in which a student reported having "four Tim Bits [mini-doughnuts] and a Coke for dinner" the previous evening. Due to a range of pressures including economic and time, parents are unable to provide their kids with healthy food. This program helps fill that need. As Jordan explains, "A lot of kids come from rural areas and in the North. A lot of parents don't have a chance to make a hot breakfast. Nut bans mean kids don't even have peanut butter and toast on the bus, their families are time crunched". Begun as a joint literacy and school meal program, funding is accessed from school board resources and community donations. Books and newspapers are available to share during and after the meal. Jordan said kids are often checking sports scores in the newspaper, and some kids will take the chance to read.

Up to 35 kids are fed breakfast twice a week. On one day they are served a hot meal including ham or bacon, eggs or pancakes, and oatmeal. On the other day kids choose from granola and yogurt. Fresh fruit or baked apples are always available. Jordan encourages kids to try foods they may not get at home. She developed one recipe where she uses pureed dates with a very small amount of brown sugar to sweeten cored apples. Served with frozen local yogurt, the kids often ask when they are having apple pie and ice cream again. Jordan is able to combine her creative flair in the kitchen and her passion for local, quality food to enhance the program.

She is also innovative when staffing the program. The breakfast program is offered in a primary school that is across the street from a high school. Jordan is able to enlist the help of high school student volunteers who get credit for their volunteer hours for their high school diplomas[4] while they learn to prepare healthy meals from fresh, local ingredients. With the program in its second year, teachers see the value in terms of more attentive students and are now helping out as well. Jordan explained that the act of eating and sharing breakfast is helping to forge new relationships between students and their teachers. Initiatives Jordan hopes to introduce this year include more shopping at the farmers' market, a cookbook, and there are plans to develop a workshop to introduce the idea to other schools in the region. The Breakfast Program provides an important service to its community, teaching children about healthy food and connecting local farmers with the school.

Local Food Plus (LFP) is another initiative that connects farmers with institutional buyers (Friedmann 2007). Originally focused in Ontario, LFP has taken its vision to other parts of Canada. LFPs role is to certify producers as meeting local production, ecological, social, animal welfare and fuel efficiency standards so institutional buyers can assure their clients that they are buying food that is community-friendly. Since negotiating ecological and social standards into food service agreements with the University of Toronto, LFP has expanded its operations into western Canada. With over 60,000 students on the University of Toronto downtown campus, the challenge is meeting the demand. LFP creates links in the community as well as,

---

4 To graduate from an Ontario high school, each student must do 40 hours of community volunteer work.

...a collaborative and flexible model of standards and verification that gives ladders to farmers and corporations to scale up local supply chains for sustainably grown products. (Friedmann 2007, 9)

The goal is to "reverse conventional incentives, and encourages regional links... to balance the scale in favour of local ecological farmers" (ibid, 11). The rapid expansion of LFP speaks to the unmet demand they are filling. From its start in 2005, the company now employs nine people and has moved into the food retail and restaurant sectors. Their goal is,

> ...to develop local sustainable food systems that meet the needs of a range of stakeholders - from farmers and processors looking for a reliable income stream, to institutions committed to purchase locally and ethically, to consumers interested in the health and social benefits of buying and eating food that is grown locally in a socially and environmentally responsible way. (Local Food Plus 2007)

Of these examples, all of these school-based projects use food as a vehicle for building new relationships. In some cases, children are learning more about how their food grows, in other cases, children are learning about healthier eating options. In all cases, eaters are reconnecting with farmers and helping to support and grow food communities that are socially and environmentally responsible.

Taking it to the Street

Other relationships are evolving that match community needs with producer capabilities to create new small-scale infrastructure. Two programs will be discussed as they each address interesting community challenges in innovative ways. The Veggie Mobile was launched in April of 2007 in three underserved New York state communities. This fruit and vegetable store on wheels brings fresh produce to disadvantaged communities that otherwise do not have reasonable access to fresh produce (Wrigley et al. 2002), "The markets' mission: to make healthy foods more affordable and more accessible to low-income residents by selling directly to the people, year-round at wholesale cost" (Capital District Community Garden 2007). The vehicle – a refrigerated cube truck with movable shelves for produce – runs on solar energy and biodiesel. The project serves seniors' centers, public housing projects and densely populated areas that lack access to fresh, healthy food. One day a week the Veggie Mobile offers a tasting program when housing project residents can sample different fruits and vegetables and take samples home to try out. The project is funded in part through the New York State Department of Health's Hunger Prevention and Nutrition Program as well as by donation.

A second project of note is the Sunshine Garden in Toronto. This initiative combined community gardening sponsored by FoodShare Toronto and connects with the Centre for Addiction and Mental Health (CAMH). FoodShare offers programs that improve food access for all Toronto citizens. According to their philosophy statement,

All of our projects are based on the premise that it's not just any food that we're talking about. We try to promote an awareness that fresh, whole foods are key to health, well-being and disease prevention, and to illustrate this principle through all our programs.

How people get their food is also important. Food distribution systems that involve communities and help to create neighborhood leaders have a great potential to enhance individual and community empowerment, by leading people to feel that they have some control over this very basic part of their lives. (FoodShare Toronto 2007a)

The Sunshine Garden was this philosophy in action. In operation from 2002 to 2006, outpatients from the CAMH worked with CAMH job coaches and FoodShare urban agriculture experts to learn about urban market gardening. They were paid to work in the program from May to October by the United Way. Produce from the garden was sold at the Sunshine Garden's own market two days a week, at a local farmers' market on Saturday and through FoodShare's Good Food Box.[5] The program was so successful that many patients reduced their medications and had a greater sense of accomplishment. According to one participant, "It really feels good to find something that needs to be done and then finish the job, knowing that you did it well" (FoodShare, CAMH n.d.). In a report on the United Way web site, the program, featured as one of the more outstanding successes, so that the participants experienced,

> ...less hospitalization, shorter stays when hospitalization does occur and improved relationships with families and peers. In 2002, only two of 60 Sunshine Garden participants became ill and had to end their participation in the program. Hospital staff say this is the first program to address the particular needs of their clients. (United Way 2007)

The garden covered 7,000 square feet and provided valuable links between the hospital, the gardeners and the local community,

> One key link between the participants and the Parkdale community members is the variety of ethnically diverse vegetables that are grown in the garden. Many of the immigrant community members are happily surprised when they see vegetables from their food culture grown locally and available fresh. These community members are eager to share their growing and cooking knowledge, providing an indispensable resource for us. (FoodShare, CAMH n.d.)

The Veggie Mobile and the Sunshine Garden capture the essence of inclusive, socially just, community food. By matching community needs and innovative use of resources, they both use food as a lens for learning about healthy eating and as an opportunity to 'enhance individual and community empowerment'.

---

5 FoodShare's Good Food Box program delivers 4,000 boxes around Toronto every month,. "The Good Food Box makes top-quality, fresh food available in a way that does not stigmatize people, fosters community development and promotes healthy eating" (FoodShare 2007b).

Building Mutual Food Systems: From Fear to Hope to Celebration

> True, radical change – a country full of people who eat food that is good for them, good
> for the people who grow it and good for the earth – is simply not coming fast enough. ...
> A revolution in how we eat means respecting food and the people who produce it.... every
> aspect of this revolution, be it related to agricultural policy, the environment or obesity,
> must begin with a plate of lovely, locally produced food and work backward from there.
> (Alice Waters in Severson 2007)

As Alice Waters remarked in this fall 2007 New York Times interview, part of the answer to the question, what do we do, depends on defining the goal – that is, what do people want to achieve? Merging the case studies presented in the previous section with: 1) the thinking about alternative versus oppositional food systems; 2) the role of agency, community and influence; and, 3) the new role for civil society, some common themes emerge. First, to recast the food system and allow people along the food chain to recapture their power, we need to engage in big picture, or systems, thinking. That is, people need to envision the kind of food they want, define the elements from field to fork that need to be sustained, and then develop a vision to make this happen.

Community, cooperation, and trust are key ingredients in building this new food system. In his latest book *Blessed Unrest*, Paul Hawken captures the groundswell of change that is emerging from the grassroots activism. As part of his book project he developed an online resource called WiserEarth that provides links to over 100,000 organizations around the world. Founded on sustainability principles and the power of the civil commons (Sumner 2005, McMurtry 2002), the organizations and people listed are all involved in moving "toward a just and sustainable world created by community" (WiserEarth 2007). Wiser Earth "serves the people who are transforming the world. It is a community directory and networking forum that maps and connects non-governmental organizations and individuals addressing the central issues of our day." (Wiser Earth 2007) Part of the solution for food systems is to scale down and go local. This, of course, is what everyone in farmers' markets, CSAs and other direct buy programmes knows.

As alluded to earlier, while there are any number of examples that demonstrate the need and pressure for the system to change, policy can be crafted to enable the evolution rather than compensate for, and prop up the IFS which seems to be failing the farmers at one end, and the consumers at the other. These examples represent public desire rather than effective movement toward a healthier – in all respects – food system. In other words, policy could shift from supporting the corporate will toward supporting the will of the host of actors including the people in their various roles and the environment. However, as Morgan et al. caution,

> If AFNs (Alternative Food Networks) are ever to become more than the marginal
> ecological spaces they are today, they will have to engage with, and draw support from,
> the multilevel governance system that regulates the agri-food system. Mobilizing political
> support for the cause of localized food chains is therefore far more than a local matter.
> (Morgan et al. 2006, 192)

According to Morgan and his colleagues, a three-pronged strategy is needed to recast these food relations so that agency, community, and influence are all activated (Halkier 2004). First, a territorial and not a sectoral approach in needed to integrate agriculture with other elements. The silos need to come down and we need to integrate all aspects of food systems to realize the multiple dividends that can accrue from healthy food communities. Second, decisions made about food systems need to be founded in subsidiarity, that is decisions should be made as low down the governmental hierarchy as possible. And third, to make this effective and relevant, consultation is needed to empower people as part of the process and to ensure that reflexivity is built into the process. Supportive policy at the macro level needs to enable responsive, relevant and robust food systems at the local level.

But, scaling down does not at all imply that we absolutely abandon top-down or scaled up policy (Morgan et al 2006; Sassatelli 2006). On the contrary – it is in fact also necessary for national governments and international bodies to become involved in promoting more localized food procurement policies as well as national and international health and labeling laws that facilitate healthier eating and food re-localization (Jackson 2004; Morgan and Morley 2003; Barling et al. 2002). National, regional, state or provincial government and non-governmental organizations can act as resource repositories and sources of funding, gathering expertise to help alternative food systems get established, standardized and labeled (Ilbery et al. 2005). Caution and vigilance are needed though to ensure that the institutions supporting inclusive activity are encouraged (Dupuis and Goodman 2005; Hinrichs 2000, 2003; Winter 2003). Regulation to foster sustainable food production is also needed. As Buller and Morris observe in their paper on market-oriented initiatives (MOIs) for environmentally sustainable food production,

> What we are arguably seeing is a reversal of the traditional division of responsibilities to a new situation in which public policy increasingly plays the role of facilitator through support schemes and payments while market forces play a greater role in regulation. As market initiatives gain in currency and, as environmental rules become increasingly built into them, then the privatization of regulation (or coregulation) that this represents raises a number of questions concerning the attainment of public policy objectives, their effects on competition in the food industry, and the division of power and the distribution of benefits in food supply chains. (Buller and Morris 2004, 1080)

This recalls the shifting regulations and areas of influence reported in Chapter 2. Both examples raise important cautions about the role of the private versus the public interests in setting the tone for regulation. Strides have been made by the UK in this direction and North Americans can learn from their example. Through the creation of progressive governance models, the UK has separated the public policy realm into more appropriate bodies for the protection, regulation and support for food, health, agriculture and commerce. At the same time, linkages have been established to permit coordination and collaboration between more branches of the government with the public and private sectors.

The evidence points to the value in multiple-scaled public education policies as important for re-framing the food system in North America. These revolutions in food systems provide a vehicle for public education that feeds change on a broader scale, and could contribute to a sea-change in the North American approach to eating and food emerging from the local scale. Public education needs to emphasize the 'difference' (Ilbery et al. 2005) so that consumers understand the benefits of buying local food for the long term – healthier food, animals and people; more robust and resilient local economies; and improved local ecologies. By building on the re-localization of food, public education programs can provoke a lasting socio-cultural shift from food fear to food celebration. In keeping with DuPuis and Goodman (2005) to move in a truly alternative direction the alternative food system needs to move away from fear and embrace democratic, reflexive localism. Multi-scaled changes from institutional change and public education can address problems head on through a respectful, inclusive process of food system reform.

Perhaps the only way to fix the system is to regain our contact with nature and understand our food system again. Given current constraints on the nature, scale and structure of current North American food systems, changing the way food is provided requires fundamental institutional and structural change at multiple dimensions and scales. This demands institutional change that in part can help shift societal patterns away from the dominance of food fears toward food as a culture of celebration and pleasure (Murdoch 2006).

The Italian example offers hope in this regard. While there are deep roots to food cultures across Italy, a closer examination shows us that institutions shape, facilitate and perpetuate a slow food versus a fast food culture (Gertler 2005; Morgan et al. 2006). Morgan et al. (2006) point out Italians have institutions and educational programs to ensure that slow food remains an integral, deliberate part of their culture. As the example of school meals in Rome makes clear, Italian food relations are constantly being renewed and reiterated. These efforts offer encouragement for other countries that a different food culture can be created, perpetuated and sustained.

There is also the opportunity to broaden the scope of food theory to engage more with ecological perspectives (Goodman 1999; Murdoch 2006). The validation of nature as an actor in food networks could shift our thinking from conceptualizing food as a commodity to thinking about food as an essential part of our culture, to be celebrated and protected. In this way we offer a hopeful vision for robust alternative food systems.

Recalling Alice Waters' comment about defining goals and working backwards, and combining this with the work of academics in the area of food studies, we discover that to create a new food system, the strategy is served better by envisioning the end goal and working towards it (Table 8.1) (Morgan et al. 2006; Allen 2004; Beus and Dunlap 1990). In other words, we decide where we want to be and work back and forth along the food chain to achieve those goals.

The rationale Freidberg uses as she advocates for "ethics of care" for African countries is also relevant for local communities in developed countries. In interrogating how to move the consumer to more local food systems she points out we need to recognize most communities are 'limited' in their capacity to create vibrant local food economies (2004a, 219). Although Friedberg is talking about globalization, we frame the change envisioned here similarly so that if change, "…is to proceed humanely, it will depend not on codes but on knowledgeable humans,

Table 8.1   Example goals and actions for building a reflexive, mutual food system

Goals for a mutual food system	Action
Provide farmers and workers from field to fork with a living wage	Develop policy and funding models to ensure farmers are paid a decent living wage, compensate farmers for environmental stewardship (e.g. pay farmers when they manage their land, air and water sustainably) Provide extension services to farmers to support them as they learn ecological farming methods Provide meaningful succession planning to preserve farmland and allow farmers to retire comfortably
Food workers	Ensure farm workers are protected by union or worker protection laws
Local sourcing, closed loop	Provide infrastructure (e.g. indoor farmers' markets, local processing facilities and distribution co-ops) that will encourage farmers to sell locally, will stabilize supply/ demand
Provide nutritious food to eaters	Determine the capacity of local foodsheds to produce healthy food from both urban and rural sources and get this food to people
Ensure animals are treated humanely	Develop and enforce humane animal standards, encourage range fed animal rearing
Food appreciation education	Education about food production, preparation and appreciation through community and school initiatives
Policy Mechanisms	Linked up departments and ministries, or a Ministry of Food per Brazil Supportive policy at macro scales that allow the micro to develop appropriate food systems at the community level

who in the face of their differences and uncertainties, can put aside their anxieties and trust each other" (Friedberg 2004a, 222).

Fear of food is a signal that something is wrong. We should not allow our societies to become inured to chronic worry, dissatisfaction, and ambivalence. From a regulation theory perspective, we are in a period of instability, and now is the time to restructure the food system so it is sustainable and community based. We have the capacity to establish the essential ingredients of a healthy food system, and to actively decide what a healthy population needs and wants. If our policy and our actions at different levels prevent our realization of these desires, then we are costing ourselves our health and our wealth. By building a community-centered food system based on principles of mutual cooperation, we capture the opportunity to embed the celebration of food into our communities and lives.

References

Aglietta, M. (1987), *A Theory of Capitalist Regulation: The US Experience* (London: Verso).

Agrafood Biotech (2006), 'Biogemma to quit Cambridge' 177: 29/5/06, 16.

Agriculture and Agri-Food Canada (AAFC) (2006), 'Agriculture and Agri-Food Canada Science and Innovation Strategy', (updated 03 July 2007) <http://www. agr.gc.ca/sci/cons/index e.php?s1=strat&page=intro>, accessed 3 September 2006.

—— (2006a), 'The Next Generation of Agriculture and Agri-Food Policy – A Federal, Provincial and Territorial Initiative', (updated 14 August 2007) <www. agr.gc.ca/nextgen>, accessed 11 February 2007.

—— (2006b), 'The Next Generation of Agriculture and Agri-Food Policy – Economic Backgrounder: Benefits and challenges of global markets', (updated 14 August 2007) <www.agr.gc.ca/ nextgen>, accessed 11 February 2007.

—— (2006c), 'The Next Generation of Agriculture and Agri-Food Policy – Environment under the Next Generation of Agriculture and Agri-Food Policy Development: A Discussion Paper', (updated 14 August 2007) <www.agr.gc.ca/ nextgen>, accessed 11 February 2007.

—— (2007), 'Agri-Food Trade Service. Trade Statistics – Canadian Trade Data by Country' <http://atn-riae.agr.ca/stats/AllCountries_country_m_e.pdf>, accessed 14 October 2007.

Allen, P. (2004), *Together at the Table: Sustainability and Sustenance in the American Agri-food System* (University Park, PA: Pennsylvania State University Press).

—— et al. (2003), 'Shifting Plates in the Agri-food Landscape: The Tectonics of Alternative Agri-food Initiatives in California', *Journal of Rural Studies.* 19:1, 61–75.

Amin, A. and Wilkinson, F. (1999), 'Learning, Proximity and Industrial Performance: An Introduction', *Cambridge Journal of Economics* 23:2, 121–125.

Anderson, R. and N. Gallini (eds) (1998), *Competition Policy and Intellectual Property Rights in the Knowledge-Based Economy, The Industry Canada Research Series* (Calgary: University of Calgary Press).

Appleby, M. et al. (2003), 'What Price Cheap Food?', *Journal of Agricultural and Environmental Ethics* 16: 395–408.

The Associated Press (2007), 'Flooding Causes Millions in Damage in Southwestern Wisconsin' (posted 20 August 2007) <http://www.greenbaypressgazette.com/ apps/pbcs.dll/article?AID=/20070820/GPG0101/70820012/1207/GPGnews>, accessed 20 September 2007.

Atkinson-Grosjean, J. et al. (2001), 'Canadian Science Policy and Public Research Organizations in the 20th Century', <http://www.oise.utoronto.ca/depts/tps/ TPS1017/FinancingResearch/Canadian%20Science%20Policy.pdf>, accessed 29 March 2007.

Aubrun, A. et al. (2005), 'Not While I'm Eating', in *Perceptions of the U.S. Food System: What and How Americans Think about Their Food*. Report prepared for the W.K. Kellogg Foundation: 1–29.

Barling, D, et al. (2002), 'Joined-up Food Policy? The Trials of Governance, Public Policy and the Food System', *Social Policy and Administration* 36:6, 556: 574.

Beck, U. (1992), *Risk Society: Towards a New Modernity* (London: Sage Publications).

Bell, D. and Valentine, G. (1997), *Consuming Geographies: We Are Where We Eat* (New York: Routledge).

Beliveau, R. et al. (2006), *Foods That Fight Cancer: Preventing Cancer through Diet* (Toronto: McClelland & Stewart).

Bergler, R (1999), 'The Effects of Commercial Advertising on Children', *Commercial Communications*, January, 41–48.

Better Farming (2001a), 'Boundary', *Syngenta*, March: 91.

Better Farming (2001b), 'Converge', *Aventis*, March: 41.

Beus, C. and Dunlap, R. (1990), 'Conventional versus Alternative Agriculture: The Paradigmatic Roots of the Debate', *Rural Sociology,* 55: 4, 590–616.

Bhaskar, R. (1979), *The Possibility of Naturalism* (Hemel Hempstead: Harvester Wheatsheaf).

Bird, J. et al. (eds) (1997), *Mapping the Future: Local Cultures, Global Change* (New York: Routledge).

Birger, J. (2007), 'The Great Gold Corn Rush', *Fortune Magazine*, 30 March.

Black, M. (1985), *Food and Cooking in 19th Century Britain, History and Recipes* (London: English Heritage).

Blay-Palmer, A. (2003), 'Growing Innovation: Agro-biotechnology and Organics Compared', PhD Thesis, Department of Geography, University of Waterloo, Waterloo, Ontario.

—— (2007), 'Who is Minding the Store? Innovation Strategy, the Social Good and Agro-biotechnology Research in Canada', *Canadian Journal of Regional Science*. 30:1, 39-56.

—— and Donald, B. (2006), 'A Tale of Three Tomatoes: The Rise of Toronto's New Food Economy', *Economic Geography* 82:4, 383–399.

—— and Donald, B. (2007), 'Manufacturing Fear: The Role of Food Processors and Retailers in Constructing Alternative Food Geographies', in Maye et al. (eds).

—— et al. (2006), 'Sustainable communities: Building Local Foodshed Capacity Through Improved Farm to Fork Links', report prepared for Frontenac and Lennox-Addington Community Futures Development Corporations, <http:// www.frontenaccfdc.com/asset.cfm?ref=as808d>, accessed 28 October 2007.

Block, F. (2001), 'Introduction', in K. Polanyi.

Bobrow-Strain, A. (2005), *Since Sliced Bread: Purity, Hygiene and the Making of Modern Bread*, paper presented at the American Association of Geographers, Denver.

—— (2006), 'Wonder Bodies: War and the Body Politics of Diet', paper presented at the *Association of American Geographers Annual Meeting*, Chicago, IL,10 March 2006.

Bordt, M. and Read, C. (1999), 'Survey of Intellectual Property Commercialization in the Higher Education Sector', *1998,* Science and Technology Redesign Project, Statistics Canada, 88F0006XPB No. 01, Ottawa.

Bostrom, A. (2005), 'Digesting Public Opinion', in *Perceptions of the U.S. Food System: What and How Americans Think about Their Food*, report prepared for the W.K. Kellogg Foundation, 31–57.

Boudreau, J-A. (2007), 'Fear, the City and Political Mobilization: An International Workshop, Canada Research Chair in the City and Issues of Insecurity: Discussion material', Montréal, 16–17 April.

Bowler, I. (2002), 'Developing Sustainable Agriculture' *Geography* 87, 205–212.

Boyer, R. (1990), The Regulation School: a Critical Introduction (New York: Columbia University).

Braun, B. (2007), 'Biopolitics and the Molecularization of Life', *Cultural Geographies* 14, 6–28.

Brenner, N. and Theodore, N. (2002), 'Cities and the Geographies of "Actually Existing Neoliberalism"', *Antipode* 34: 3, 349–37.

Briggs, A. (1988), 'Marks and Spencer P.L.C. International Directory of Company Histories', Volume 5, 124–126.

British Broadcasting Corporation (BBC) 2002. 'Health: Peanut Allergy on the increase' 18 November 2002 <http://news.bbc.co.uk/2/hi/health/2487769.stm>, accessed 6 October 2007.

—— (2007), 'Tighter Rules on Food Ads Urged: Teaching, Health and Consumer Groups Have Written to the Government Urging it to Tighten Up the Rules on Food Adverts that Target Children', (published online 27 April 2007) <http://news.bbc.co.uk/2/hi/uk_news/6601465.stm>, accessed 21 May 2007.

—— (2007a), 'BBC Food News.' <http://www.bbc.co.uk/food/news_and_events/foodnews_september.shtml>, accessed 2 October 2007.

Britvic (2006), 'Soft drinks category report', <http://www.britvic.com/PDF/BRITVIC_CATEGORY_REPORT_LR.PDF>, accessed 21 May 2007.

—— (2007), 'Britvic: FTSE sector: Beverages', (updated 19 February 2007) <http://www.britvic.com/PDF/BritvicInBrief.pdf>, accessed 21 May 2007.

British Ornithologists' Union (BOU), 2006, 'Avian Influenza and Other Bird Diseases', Press Release, December 01, (updated 30 August 2007), <http://www.bou.org.uk/>, accessed 1 August 2007.

—— (2007), 'Avian Influenza: Migratory Birds: Innocent Scapegoats for the Dispersals of the H5N1 Virus', March 22, (updated 30 August 2007) <http://www.bou.org.uk/>, accessed 1 August 2007.

Buller, H. and Morris, C. (2004), 'Growing goods: the market, the state and sustainable food production' *Environment and Planning A* 36, 1065–1084.

Bunge (2007), 'Bunge', <http://www.bunge.com/corporate-home.html>, (home page), accessed 30 May 2007.

Bunnell, T. and Coe, N. (2001), 'Spaces and Scales of Innovation', *Progress in Human Geography* 25:4, 569–589.

Burnett, J. (1989), *Plenty and Want A Social History of Food in England from 1815 to the Present Day*, 3rd ed., (London: Routledge).

Burt, S. et al. (2002), 'Retail Internationalization and Retail Failure: Issues from the Case of Marks and Spencer', *International Review of Retail, Distribution and Consumer Research* 12:2, 191–219.

Busch, L. et al. (1991), *Plants, Power and Profit: Social, Economic and Ethical Consequences of the New Biotechnologies* (Cambridge, M.A.: Basil Blackwell).

Byrne, P. (2007), 'Labeling of Genetically Engineered Foods'. Colorado State University Extension – Nutrition Services, <http://www.ext.colostate.edu/pubs/foodnut/09371.html> accessed 27 September 2007.

Callon, M. (1986), 'Some Elements of a Sociology of Translation: Domestication of the Scallops and the Fishermen of St Brieuc Bay', in J. Law, (ed.).

Canadian Broadcasting Corporation (CBC) (1948), 'Housewives Save with Margarine', <http://archives.cbc.ca/IDC-1-69-405-2323-10/life_society/margarine_ban_lifted/> accessed 23 April 2007.

—— (2001), '*Environmental, Consumer Groups Demand Mandatory GM Food Labels*', <http://www.cbc.ca/story/news/national/2003/03/13/Consumers/gmlabeling_030313.html>, accessed 3 September 2006.

—— (2006a), 'Mad Cow in Canada: The Science and the Story', CBC News Online, 24 August 2006, <http://www.cbc.ca/news/background/madcow/index.html>, accessed 5 July 2007.

—— (2006b), 'Timeline of BSE in Canada and the U.S', CBC News In Depth Online 23 October , <http://www.cbc.ca/news/background/madcow/timeline.html>, accessed 5 July 2007.

—— (2007a), 'Milk Safe because it's Treated, Producers Point Out' CBC News, <http://www.cbc.ca/health/story/2007/10/05/qc-milksafe1005.html?ref=rss> , accessed 25 October 2007.

—— (2007b), 'Food Safety: 'A Safer Food Supply', Buyer Beware,' CBC News In Depth, <http://www.cbc.ca/news/background/foodsafety/>, 5 July 2007.

Canadian Food Inspection Agency (CFIA) (2007a), 'Enhanced Health Protection from BSE' (updated January 30 2007) <http://www.inspection.gc.ca/english/anima/heasan/disemala/bseesb/enhren/publie.shtml>, accessed 5 July 2007.

—— (2007b), 'About the CFIA', (updated 28 October 2006) <http://www.inspection.gc.ca/english/agen/agene.shtml>, accessed 5 July 2007.

Canadian Poultry Industry Council (2005), 'No Need for Bird Flu to Become Bird Fear – Ontario Agriculture's Perspective on Avian Influenza', <http://www.poultryindustrycouncil.ca/pdf/ai/Avian%20flu%20Op%20Ed.pdf>, accessed 28 October 2007.

Cannon, G. (1986), 'Reaction to Food Additives', *British Medical Journal* 292: 10 May, 1275.

Capital District Community Gardens (2007), Programs: Veggie Mobile, <http://www.cdcg.org/VeggieMobile.html>, accessed 23 September 2007.

Carolan, M. (2006), 'Risk, Trust, and "The Beyond" of the Environment: A Brief Look at the Recent Case of Mad Cow Disease in the United States', *Environmental Values* 15, 233–252.

Carr, S. (2001), 'Ethical and Value-Based Aspects of the European Commission's Precautionary Principle', *Journal of Agricultural and Environmental Ethics* 15, 31–38.

Carson, R. (1962), *Silent Spring* (Boston: Houghton Mifflin).

Cassier, M. (2002), 'Private Property, Collective Property and Public Property in the Age of Genomics', *International Social Science Journal* 54:171, 83–98.

Castree, N. (2002) 'False Antitheses? Marxism, Nature and Actor-Networks', *Antipode* 34, 111–146.

Center for Disease Control and Prevention (2007), 'BSE (Bovine Spongiform Encephalopathy, or Mad Cow Disease)', <http://www.cdc.gov/ncidod/dvrd/bse/>, accessed 03 October 2007.

Center for Food Safety (2007), *rBGH/Hormones* <http://www.centerforfoodsafety. org/rbgh_hormo.cfm>, accessed 12 April 2007.

Center for Infectious Disease Research and Policy (CIDRAP) (2006), 'USDA to Cut back BSE testing'. Academic Health Center, University of Minnesota. <http:// www.cidrap.umn.edu/cidrap/content/other/bse/news/jul2006bse.html>, accessed 3 October 2007.

Chettero, N. (2005), TransFair USA Joins Oxfam in Welcoming McDonald's Rollout of Fair Trade Certified(TM) Coffee. Fair Trade Certified. <http://www. transfairusa.org/content/about/pr_051031.php>, accessed 25 November 2006.

Clausheide, G. and Courtenay-Hall, P. (2007), Personal communications and notes, Prince Edward Island, July and August.

Coleman, E. et al. (2003), 'Agribusiness Group Paper: Agribusiness Defined', The Industrial College of the Armed Forces National Defense University Fort McNair Washington, DC 20319-5062 <http://stinet.dtic.mil/cgi-bin/GetTRDoc?AD=AD A425302&Location=U2&doc=GetTRDoc.pdf>, accessed 28 September 2007.

The Court of Appeal for Saskatchewan (2007), 'Docket: 1148 between Larry Hoffman. L.B.Hoffman Farms Inc., and Dale Beaudoin', Appellant and Monsanto Canada Inc. and Bayer Cropscience Inc., Respondent, Citation: 2007 SKCA 47 Date: 20070502. <http://www.saskorganic.com/oapf/pdf/2007SKCA047.pdf>, accessed 25 May 2007.

Cover Concepts (2003), 'Cover Concepts – Free Stuff for Kids', <http://www. coverconcepts.com/>, accessed 21 May 2007.

Cox, K. (ed.) (1997), *Spaces of Globalization – Reasserting the Power of the Local* (New York: Guilford Press)

Cox, K. (1998), 'Spaces of Dependence, Spaces of Engagement and the Politics of Scale, or: Looking for Local Politics', *Political Geography* 17:1, 1–23.

Cronon, W. (1996), 'Forward', in Cronon (ed.).

Cronon, W. (ed.) (1996), *Uncommon Ground: Rethinking the Human Place in Nature* (New York: Norton and Company).

Culliton, B. (1977), 'Harvard and Monsanto: the $23-Million Alliance' *Science* 195, 759–763.

Cussler, M. (1952), *Twixt the Cup and the Lip: Psychological and Socio-cultural Factors Affecting Food Habits* (New York: Twayne Publishers).

Dairy Farmers of Ontario (DFO) (2007), Quota Exchange Information, <http://www.milk.org/Corporate/View.aspx?Content=Farmers/QuotaExchangeInformation>, accessed 29 October 2007.

Dale, S. (2005), *Candy from Strangers: Kids and Consumer Culture* (Vancouver: New Star Books).

David, P. (ed.) (1993), *Proceedings of the World Bank Annual Conference on Development Economics 1992* (Washington, DC: World Bank).

—— (2001), *A Tragedy of the Public Knowledge 'Commons': Global Science, Intellectual Property, and the Digital Technology Boomerang*, (Oxford and Stanford University) <http://www.oiprc.ox.ac.uk/EJWP0400.pdf>, accessed 22 October 2007.

DECIMA (2004), 'Canadian Consumer Poll: Eating GM Food', <http://www.cbc.ca/consumers/market/files/food/gmfood/index2.html>, accessed 3 September 2006.

Department for Environment, Food and Rural Affairs (DEFRA) (2007), 'Outbreak Of Highly Pathogenic H5N1 Avian Influenza In Suffolk In January 2007', *A Report Of The Epidemiological Findings By The National Emergency Epidemiology Group, DEFRA*, 5 April 2007.

Desai, A. (1993), 'Comment on 'Knowledge, Property, and the System Dynamics of Technological Change', in David, P. (ed.).

Dimitri, C. and Oberholtzer, L. (2005), *Market-Led Growth vs. Government-Facilitated Growth: Development of the U.S. and EU Organic Agricultural Sectors*, USDA Outlook Report No. WRS0505.

Doern, B. (1999), *Global Change and Intellectual Property Agencies, An Institutional Perspective* (New York: Pinter).

Donald, B. and Blay-Palmer, A. (2006), 'The Urban Food Economy: Cultural Consumption for the Urban Elite or Social Inclusion Opportunity?' *Environment and Planning A* 38:10, 1901–1920.

Dow Jones International News. (2005),'Rumsfeld Garners $1M In Recent Run-Up of Gilead Stock', 2 November, 11:41.

Draper, A. and Green, J. (2002), 'Food Safety and Consumers: Constructions of Risk and Choice', *Social Policy and Administration* 36:6, 610–625.

Drottberger, A. (2005), 'The Relationship Between Vegetable Growers and the Supermarkets. Undergraduate thesis. Sveriges Lantbruksuniversitet. ex-epsilon. slu.se/archive/00000647/.

Drummond, J. and Wilbraham, A. (1958), *The Englishman's Food* (London: Jonathan Cape).

DuPont. 2002, 'DuPont Crop Protection, Battalion™', <http://www.dupont.ca/ag/product/index.cfm?inylangid=1&CropSection=Corn&ItemID=264>, accessed 24 May 2007.

Dupré, R. (1999), 'If It's Yellow, It Must Be Butter: Margarine Regulation in North America Since 1886', *The Journal of Economic History* 59:2, 353–371.

DuPuis, M. and Goodman, D. (2005), 'Should We "Go Home" To Eat?: Towards a Reflexive Politics of Localism', *Journal of Rural Studies* 21, 359–371.

Earthbound Farm Organic 2007. 'About Us'. <http://www.ebfarm.com/AboutUs/OurMission.aspx>, accessed 13 October 2007.

Eder, K. (1996), *The Social Construction of Nature* (London: Sage Publications Limited).

The Edible Schoolyard (2006a), *About Us: The Edible Schoolyard*, Martin Luther King, Jr. Middle School, Berkeley California, <http://www.edibleschoolyard.org/about.html> accessed 22 September 2007.

The Edible Schoolyard (2006b), *The People: The Edible Schoolyard*, Martin Luther King, Jr. Middle School, Berkeley California, <http://www.edibleschoolyard.org/ppl_mlk.html>, accessed 22 September 2007.

Eisen, J. (ed.) (1994), 'Ideas: Biotechnology's Harvest', (Toronto: Canadian Broadcasting Corporation) (transcript from Ideas program 15, 22 March 1994).

Engle, M. (2004), 'Regulating Food Advertising to Children: An Historical Perspective', *Institute of Medicine Meeting on Food Marketing and the Diets of Children and Youth*, Washington, D.C. October 14 2004.

Enticott, G. (2003), 'Risking the Rural: Nature, Morality and the Consumption of Unpasteurised Milk', *Journal of Rural Studies* 19 411–424.

Environics Research Group (2001), 'Cross Country Consumer Poll: Genetic Engineered Food Safety and Labelling'. http://www.canadians.org/publications/CP/2000/CP_Spring_00.pdf, accessed 22 October 2007.

European Commission Health & Consumer Protection Directorate-General (2002), 'Opinions of the Scientific Committee on Veterinary Measures Relating to Public Health On Review of Previous SCVPH opinions of 30 April 1999 and 3 May 2000 on the Potential Risks to Human Health from Hormone Residues in Bovine Meat and Meat Products', <http://ec.europa.eu/food/fs/sc/scv/out50_en.pdf>, accessed 12 April 2007.

Evans, P. (1995), *Embedded Autonomy: States and Industrial Transformation* (Princeton: Princeton University Press).

Evenson, R. and Gollin, D. (2003), 'Assessing the Impact of the Green Revolution, 1960 to 2000', *Science* 300: 5620, 758–762.

Farb, P. and Armelagos, G. (1980), *Consuming Passions: The Anthropology of Eating* (Boston: Houghton Mifflin).

Featherstone, M. et al. (eds) (1991), *The Body: Social Processes and Cultural Theory* (London: Sage).

Federal Emergency Management Agency (FEMA) (2007), '2007 Federal Disaster Declarations: Minnesota Severe Storms and Flooding, Declared August 23 2007', <http://www.fema.gov/news/event.fema?id=8665>, accessed 27 August 2007.

Fidelman, C. (2006), 'Food Ads Making Kids Heavy, Expert Warns', *Montreal Gazette*, 27 October 2006.

Fisher, D. et al. (2001), 'Changes in Academy/Industry/State Relations in Canada: The Creation and Development of the Networks of Centres of Excellence', *Minerva* 39, 299–325.

Fischler, C. (1988), 'Food, Self and Identity', *Social Science Information* 27:2, 275–292.

Fitzsimmons, M. (1989), 'The Matter of Nature', *Antipode* 21:2, 106–120.

Fleising, U. (2002), 'The Legacy of Nuclear Risk and the Founder Effect in Biotechnology Organizations', *TRENDS in Biotechnology* 20:4, 156–1159.

Food and Agriculture Organization/ World Health Organization (2005), 'Codex Alimentarius: Food Import and Export Inspection And Certification Systems: Second edition Joint FAO/WHO Food Standards Programme Codex Alimentarius Commission' <ftp://ftp.fao.org/docrep/fao/008/y6396e/y6396e00.pdf>, accessed 25 September 2007.

—— (1999), 'Recommended International Code of Practice, General Principles of Food Hygiene, CAC/RCP 1-1969, Rev. 3 (1997), Amended 1999' <http://www.fao.org/DOCREP/005/Y1579E/y1579e02.htm>, accessed September 25, 2007.

Foodland Ontario. 2007, 'Supervising the Foodland Ontario Brand', Ontario Ministry of Food, Agriculture and Rural Affairs, (created 1 January 2002, last reviewed: 17 September 2004) <http://www.omafra.gov.on.ca/english/food/domestic/brandservices.htm>, accessed 14 January 2007.

FoodShare (2007a), 'Our Philosophy', http://www.foodshare.net/whoweare05.htm, accessed 28 October 2007.

—— (2007b), 'What is the Good Food Box?' <http://www.foodshare.net/goodfoodbox01.htm>, accessed 10 October 2007.

—— and CAMH (2007), 'The Sunshine Garden' http://www.foodshare.net/download/CAMHPromo-s.pdf, accessed 29 October 2007.

Foray, D. (1995), 'The Economics of Intellectual Property Rights and Systems of Innovation: The Persistence of National Practices Versus the New Global Model of Innovation', in Hagedoorn, (ed.).

Foucault, M. (1977), *Discipline and Punishment: the Birth of the Prison* (London: Lane).

—— (2003), *Society Must Be Defended* (New York: Picador).

Friedberg, S. (2004), *French Beans and Food Scares: Culture and Commerce in an Anxious Age* (New York: St. Martin's Press).

—— (2004a), 'The Ethical Complex of Corporate Food Power', *Environment and Planning D: Society and Space* 22, 513–531.

Friedmann, H. (1993), 'The Political Economy of Food: A Global Crisis', *New Left Review* 197, 29–57.

—— (2004), 'Eating in the Gardens of Gaia: Envisioning Polycultural Communities', in Adams, (ed.).

—— (2007), 'Scaling up: Bringing Public Institutions and Food Service Corporations into the Project for a Local, Sustainable Food System in Ontario', *Agriculture and Human Values* 24: 389–398.

Fuller, B. (1963), *Operating Manual for Spaceship Earth* (New York: E.P. Dutton & Co.)

Fulwood, C. (1996), 'Alar Report Right from the Start, But You'd Never Know It'. *Public Relations Quarterly* 41:2, 9–12.

Gabriel, Y. and Lang, T. (2006), *The Unmanageable Consumer*, 2nd Edition (London: Sage).

Gale, P. (2006), 'Review Article: BSE Risk Assessments in the UK: a Risk Tradeoff?' *Journal of Applied Microbiology* 100, 417–427.

Gallo, A. (1999) *Food Advertising in the United States. America's Eating Habits: Changes and Consequences* (Washington, DC: USDA/Economic Research Service).

Gardener, C. and Sheppard, J. (1989), *Consuming passion: the Rise of Retail Culture* (London: Unwin Hyman).

Gaskell, G. et al. (2003), 'Europeans and Biotechnology in 2002 Eurobarometer 58.0 2nd Edition, A Report to the EC Directorate General for Research from the Project 'Life Sciences in European Society', <http://www.oeaw.ac.at/ita/ebene5/HT_1176.pdf>, accessed 22 October 2007.

Gauthier-Clerc, M. et al. (2007), 'Expansion of Highly Pathogenic Avian Influenza H5N1: A Critical Review', *Ibis* 149:2, 202–214.

Gertler, M. (2005), London Calling: Five Myths of Cluster Development, *Innovation Systems Research Network Annual Meeting*, Toronto, Canada, May 5–7.

Gifford Jr, A. (1997), 'Whiskey, Margarine, and Newspapers: A Tale of Three Taxes', in Shugart (ed.).

Gilbert, E. (2005), 'The Inevitability of Integration? Neoliberal Discourse and the Proposals for a New North American Economic Space after September 11', *The Annals of the Association of American Geographers* 95:1, 202–222.

Glassner, B. (1999), *The Culture of Fear: Why Americans Are Afraid of the Wrong Things* (New York: Basic Books).

Gold, M. (2004), 'The Global Benefits of Eating Less Meat'. Report for Compassion in World Farming Trust. <http://www.ciwf.org/publications/reports/The_Global_Benefits_of_Eating_Less_Meat.pdf> accessed 3 June 2007.

Gogoi, P. (2006), Wal-Mart's Organic Offensive, BusinessWeek Online. March 29. <http://www.businessweek.com/bwdaily/dnflash/mar2006/nf20060329_6971.htm>, accessed 25 November 2006.

Goodman, D. and Redclift, M. (1991), *Refashioning Nature: Food, Ecology and Culture* (London: Routledge).

—— and Watts, M. (eds.) (1997), *Globalising Food: Agrarian Questions and Global Restructuring* (New York: Routledge).

—— (1999), 'Agro-Food Studies in the "Age of Ecology": Nature, Corporeality, Bio-politics', *Sociologia Ruralis* 39:1, 17–38.

—— (2003), 'Editorial: The Quality 'Turn' and Alternative Food Practices: Reflections and Agenda' *Journal of Rural Studies* 19: 1–7.

—— (2004), 'Rural Europe Redux? Reflections on Alternative Agro-Food Networks and Paradigm Change', *Sociologia Ruralis*. 44:1, 3–16.

Gotsch, N. and Rieder, P. (1995), 'Biodiversity, Biotechnology, and Institutions among Crops: Situation and Outlook', *Journal of Sustainable Agriculture* 5:1/2, 5–40.

Government of Australia, Department of Agriculture, Fisheries and Forestry (2006), 'BSE-Free Status Confirmed for Australia' <http://www.daffa.gov.au/about/media-centre/dept-releases/2006/test_media_releases>, accessed 7 August 2007.

GRAIN (2006), 'Fowl Play: The Poultry Industry's Central Role in the Bird Flu Crisis', *GRAIN Briefing*, February 2006, <http://www.grain.org/briefings/?id=194>, accessed 2 August 2007.

—— (2007), 'GRAIN: About Us' <http://www.grain.org/about/>, accessed 3 October 2007.

Granovetter, M. (1985), 'Economic Action and Social Structure: The Problem of Embeddedness', *American Journal of Sociology* 91:3, 481–510.

Greenberg, B. and Brand, J. (1993), 'Television news and advertising in schools: The "Channel One" controversy', *Journal of Communication* 43:1, 143–151.

Gussow, J. (2001), *Reinventing the World: Food* Video directed by David Springbett and Heather MacAndrew. Produced by Asterisk Productions Ltd. Host and narrator, Des Kennedy. Produced in association with Vision TV, distributed by Bullfrog Films.

Guthman, J. (2003), 'Fast Food/Organic Food: Reflexive Tastes and the Making of "yuppie chow"', *Social & Cultural Geography* 4:1, 45–58.

—— (2004), *Agrarian Dreams: The Paradox of Organic Farming in California* (Berkeley: University of California Press).

—— (2004b), 'The Trouble with "Organic Lite" in California: A Rejoinder to the "Conventionalisation" Debate', *Sociologia Ruralis* 44:3, 301–316.

Haenn, N. (2002), 'Nature Regimes in Southern Mexico: A History of Power and Environment', *Ethnology* 41:1, 1–26.

Hagedoorn, J. (ed.) (1995), *Technical Change and the World Economy* (Brookfield, Vermont: Edward Elgar Publishing Limited).

Halkier, B. (2004), 'Handling Food-related Risks: Political Agency and Governmentality', in Lien and Nerlich (eds).

Hallman, W. et al. (2003), 'Public Perceptions of Genetically Modified Foods: A National Study of American Knowledge and Opinion', *New Brunswick, New Jersey: Food Policy Institute*, Cook College, Rutgers – The State University of New Jersey.

Hamilton, C. et al. (2006), 'The Rendering Industry's Bio-security Contribution to Public and Animal Health' in D. Meeker (ed.) (2006) *Essential Rendering: All About The Animal By-Products Industry* (Arlington VA: Kirby Lithographic Company, Inc.) <http://www.animalprotein.org/PDF/Essential-Rendering.pdf>, accessed 2 October 2007.

Hardy, A. (1999), 'Discussion Point Food, Hygiene and the Laboratory. A Short History of Food Poisoning in Britain, circa 1850–1950', *The Society for Social History of Medicine*, 12:2, 293–311.

—— (2004), 'Salmonella: a Continuing Problem', *Postgraduate Medical Journal* 80:947, 541–545.

Harhoff, D. et al. (2001), 'Genetically Modified Food: Evaluating the Economic Risk', *Economic Policy,* October: 264–299.

Harris Ali, S. and Keil, R. (2006), 'Global Cities and the Spread of Infectious Disease: The Case of Severe Acute Respiratory Syndrome (SARS) in Toronto, Canada', *Urban Studies*, 43 (3): 491–509.

Harvey, D. (2000), *Spaces of Hope* (Berkeley: University of California Press).

—— (1989), *The Condition of Post-modernity: An Enquiry into the Origins of Cultural Change* (Cambridge, MA: Blackwell Press).

Hawken, P. (2007), *Blessed Unrest: How the Largest Movement in the World Came into Being and Why No One Saw It Coming* (New York: Viking Press).

Heller, L. (2007), 'Food safety risks stick with consumers long-term', survey, Food USANavigator.com,<http://www.foodnavigator-usa.com/news/ng.asp?id=73144>, accessed 1 August 2007.

Heller, M. and Eisenberg, R. (1998), 'Can Patents Deter Innovation? The Anticommons in Biomedical Research', *Science* 280:5364, 698–701.

Henderson, M. 2005. 'Scientists Aim to Beat Flu with Genetically Modified Chickens', *The Times* 29 October, 2005, <http://www.timesonline.co.uk/tol/news/world/article584036.ece>, accessed 3 October 2007.

Hewitt, M. (1991), 'Bio-politics and Social Policy: Foucault's Account of Welfare', in Featherstone et al. (eds).

Hickman, M. (2006), 'Was Edwina Currie Right About Salmonella in Eggs, After All?', *The Independent*, (London) 17 November <http://findarticles.com/p/articles/mi_qn4158/is_20061117/ai_n16863347>, accessed 4 July 2007.

Hinrichs, C. (2000), 'Embeddedness and local food systems: notes on two types of direct agricultural market', *Journal of Rural Studies* 16:3, 295–303.

—— (2003), 'The Practice and Politics of Food System Localization', *Journal of Rural Studies* 19, 33–45.

Holloway, L. and Kneafsey, M. (2000), 'Reading the Space of the Farmers' Market: a Preliminary Investigation from the UK', *Sociologia Ruralis* 40, 285–299.

—— et al. (2005), 'Possible Food Economies: Food Production-Consumption Arrangements and the Meaning of "Alternative"', Cultures of Consumption working paper, <http://www.consume.bbk.ac.uk/publications.html>, accessed 24 October 2007.

—— et al. (2007), 'Beyond the "Alternative" – "Conventional" Divide? Thinking Differently About Food Production-Consumption Relationships', in Maye et al. (eds).

Horbulyk, T. (1993), 'Intellectual Property Rights and Technological Innovation in Agriculture', *Technological Forecasting and Social Change* 43: 259–270.

Hudson, R. and Elwell, L. (2004), 'Avian Influenza, Lessons Learned and Moving Forward', report from the 2004 Canadian Poultry Industry Forum <http://www.agf.gov.bc.ca/avian/CPIF-Avian.pdf>, accessed 3 October 2007.

Hulme, G. (2007), 'Wal-Mart Stores Inc. 2007. In-Front with Wal-Mart', Information Week, 040, 39–42, <http://infrontwithwalmart.com/segments.aspx?t=5>, accessed 3 October 2007.

Humane Society of the United States (2006), 'The Welfare of Animals in the Broiler Chicken Industry', <http://www.hsus.org/farm/resources/research/welfare/broiler_industry.html>, accessed 3 October 2007.

Hyrd, M. (2007), Personal communication. Departments of Sociology and Geography, Queen's University, Kingston Ontario.

Ilbery, B., and Kneafsey, M. (2000), "Producer Constructions of Quality in Regional Specialty Food Production: A Case Study from South West England", *Journal of Rural Studies*, Vol. 16 No.2, pp.217–30.

—— et al. (2005), 'Product, Process and Place: An Examination of Food Marketing and Labelling Schemes in Europe and North America', *European Urban and Regional Studies* 12:2, 116–132.

—— and Maye, D. (2005a), 'Alternative (shorter) Food Supply Chains and Specialist Products in the Scottish – English Borders', *Environment and Planning A* 37, 823–844.

—— and Maye, D. (2005b), 'Food Supply Chains and Sustainability: Evidence from Specialist Food Producers and in the Scottish/ English Borders', *Land Use Policy* 22, 331–344.

Internet World Stats (2007), 'World Usage and Population', <http://www.internetworldstats.com/>, accessed 21 May 2007.

Iowa Farm Bureau, '*The Economics of the E.U. Beef Hormone Ban and Carousel Retaliation*', http://www.iowafarmbureau.com/programs/commodity/pdf/eoa7.pdf>, accessed 30 April 2007.

Jackson, T. (2004), *Motivating Sustainable Consumption: A Review of Evidence on Consumer Behaviour and Behavioural Change*, report to the Sustainable Development Research Network.

Jackson, P. et al. (2007), 'The Appropriation of "Alternative", Discourses by "Mainstream" Food Retailers', in D. Maye (eds).

Jenson, J. (1989), 'Different but not Exceptional: Canada's Permeable Fordism', *Canadian Review of Anthropology and Sociology*, 26:1, 69–93.

Jolly Time Popcorn (2004), 'JOLLY TIME Pop Corn Expands "Smart Snack" Offerings With Healthy Pop 94% Fat Free Caramel Apple Microwave Pop Corn', Press Release, <http://www.jollytime.com/newsroom/releases/caramel_apple.asp>, accessed 22 September 2007.

Kagan, R. et al. (2003), 'Prevalence of Peanut Allergy in Primary-school Children in Montreal, Canada', *Journal of Allergy and Clinical Immunology*, 112:6, 1223–1228.

Kaiser Family Foundation (2007), '*Food for Thought: Television Food Advertising to Children in the United States*', http://www.kff.org/entmedia/entmedia032807pkg.cfm, accessed 24 October 2007.

Kalra E. (2003), 'Nutraceutical – Definition and Introduction', *American Association of Pharmaceutical Scientists Journal* 5:2.

Kautsky, K. (1902), 'Socialist Agitation Among Farmers in America', *International Socialist Review* 3, 148–160. Translated by Ernest Untermann from German *Bauernagitation in Amerika*, Die Neue Zeit, 1902, Vol.XX(2): 453. http://www.marxists.org/archive/kautsky/1902/09/farmers.htm, accessed 26 September 2007.

Kelley, Hubert W. (1993), 'W.O. Atwater – Father of American Nutrition Science', *Agricultural Research Magazine*, <http://www.ars.usda.gov/is/timeline/nutrition.htm?pf=1>, accessed 11 April 2007.

Kellogg's (2007), Kellogg's Apple Jacks web site. <http://www.applejacks.com/scuba.shtml>, accessed 21 May 2007.

Kilgour, D. (2004), 'Innisfree BSE Symposium, Innisfree', personal notes, 16 October, <http://www.david-kilgour.com/mp/Innisfree.htm>, accessed 12 August 2007.

Kingston, A. (1994), *The Edible Man: Dave Nichol, President's Choice, and the making of popular taste* (Toronto: Macfarlane Walter and Ross).

Kingston Economic Development Corporation (KEDCO) (2004), Kingston Profile 2004: Economic Base, <http://business.kingstoncanada.com/communityprofile/resources/Economic.pdf>, accessed 31 July 2006.

Kirschenman, F. (2005), 'Challenges and Opportunities to Going Local'. Caledon Countryside Alliance and Toronto Food Policy Council Conference, *Farm Folk City Folk: How New Alliances Can Bring Local Food to the Table*, 12 November.

Kirwan, J. (2004), 'Alternative Strategies in the UK Agro-Food System: Interrogating the Alterity of Farmers' Markets', *Sociologia Ruralis* 44:4, 395–415.

Klint-Jensen, K. (2004), 'BSE in the UK: Why the Risk Communication Strategy Failed' *Journal of Agricultural and Environmental Ethics* 17, 405–423.

Kloppenberg, J. et al. (1996), 'Coming Into the Foodshed', *Agriculture and Human Values* 13:3, 33–41.

Kneafsey, M., et al. (2004), *Consumers and Producers: Coping with Food Anxieties through 'Reconnection'?* Cultures of Consumption and ESRC-AHRB Research Program Working Paper Series, Working Paper No. 19, <http://www.consume.bbk.ac.uk/publications.html#workingpapers>, accessed 24 October 2007.

Kneen, B. (ed.) (1993), *From Land to Mouth: Understanding the Food System* (Toronto: NC Press).

Kneen, B. (1993), *Distancing: the Logic of the Food System*, in Kneen (ed.)

Kortbech-Olesen, R. (2004), 'The Canadian Market for Organic Food and Beverages'. Report prepared for *International Trade Centre (UNCTAD-WTO)*, <http://www.intracen.org/menus/search.htm>, accessed 24 October 2007.

Kroma, M. and Flora, C. (2003), 'Greening Pesticides: A Historical Analysis of the Social Construction of Farm Chemical Advertisements', *Agriculture and Human Values* 20: 1, 21–35.

Lam, S. (2006), *Food Miles: Environmental Implications of Food Imports to the Kingston Region: Brief Summary of Findings and Comparison to Waterloo Region,* unpublished Masters Thesis: Queen's University, Kingston Ontario.

Lamine, C. (2005), 'Settling Shared Uncertainties: Local Partnerships Between Producers and Consumers', *Sociologia Ruralis* 45:4, 324–345.

Lang, T. and Heasman, M. (2004), *Food Wars: The Global Battle for Minds, Mouths and Markets* (London: Earthscan).

—— and Hallman, W. (2005), 'Who Does the Public Trust? The Case of Genetically Modified Food in the United States', *Risk Analysis* 25:5. 1241–1252.

Larner, W. and Le Heron, R. (2002), 'From Economic Globalisation to Globalising Economic Processes: Towards Post-Structural Political Economies', *Geoforum* 33:4, 415–419.

Law, J. (ed.) (1986), *Power, Action and Belief: A New Sociology of Knowledge?* (London: Routledge).

Lawson, V. (2007), 'Introduction: Geographies of Fear and Hope' *Annals of the Association of American Geographers*, 97:2, 335–337.

Le Heron, R. (2003), 'Cr(Eat)ing Food Futures: Reflections on Food Governance Issues in New Zealand's Agri-food Sector', *Journal of Rural Studies* 19:1, 111–125.

—— (2006), 'Towards Governing Spaces Sustainably: Reflections in the Context of Aukland, New Zealand', *Geoforum* 33:4, 415–419.

Lee, B. (1961), 'The Atom Secrets,' *Globe Magazine*, 28 October 1961.

Lee, R., and Leyshon, A. (2003), 'Conclusions: Re-making Geographies and the Construction of "Spaces of Hope"', in Leyshon (eds).

Lemke, T. (2001), '"The birth of bio-politics"', Michel Foucault's lecture at the College de France on neo-liberal governmentality', *Economy and Society* 30:2, 190–207.

Levenstein, H. (1993), *Paradox of Plenty: A Social History of Eating in Modern America* (New York: Oxford University Press).

—— (2003), *Revolution at the Table: The Transformation of the American Diet* (Berkeley, University of California Press).

—— (2004), *Fear of Food*. CBC Ideas Program Transcripts. Developed by Jill Eisen. Produced by Alison Moss with the assistance of Liz Nagy. Archival research by Debra Lindsay. Technical operations by Dave Field.

Leyshon, A. et al. (eds) (2003), *Alternative Economic Spaces* (London: Sage Publications).

Lien, M-A. and Nerlich, B. (eds) (2004), *The Politics of Food* (New York: Berg).

—— (2004), 'The Politics of Food: An Introduction', in Lien and Nerlich (eds).

Lipietz, A. (1995), *Green Hopes: the Future of Political Ecology* (Cambridge: Polity Press, Blackwell Publishers Ltd).

Liquori, T. (2007), 'Tutto per Qualitá: Innovation in Rome's School Meal System'. Program in Nutrition, Teachers College, Columbia University. <http://www. worldhungeryear.org/why_speaks/ws_load.asp?file=80&style=ws_print>, accessed 10 October 2007.

Local Food Plus (2007), 'Why Local Sustainable Food? The Issues.' <http://www. localflavourplus.ca/why_local_sustainable_food.htm>, accessed 10 October 2007.

Los, F. (2006) 'The Terminator' *Alternatives Journal*, 32, 3, 2006, 24–26.

Lucas, C. and Hines, C. (2006), 'Avian flu: Time to Shut the Intensive Poultry "Flu Factories"?', <http://www.carolinelucasmep.org.uk/framesets/publications. html>, accessed 28 May 2007.

Lund, B.M. (2000), 'Freezing', in Lund (eds).

—— et al. (eds) (2000), *The Microbiological Safety and Quality of Food* Vol. 1 (Gaithersburg, MD: Aspen Publishers).

Lundvall, B.A. (1993), 'User-producer Relationships, National Systems of Innovation and Internationalization', in Foray (eds).

MacRae, R., Henning, J. and Hill, S. (1993), 'Strategies to Overcome Barriers to the Development of Sustainable Agriculture in Canada: The Role of Agribusiness', *Journal of Agricultural and Environmental Ethics* 6:1, 21–51.

Manitoba Farm and Rural Stress Line (2005), 'Annual Report', <http://www. ruralstress.ca/files/MFRSL_2005.pdf>, accessed 12 August 2007.

Mann, S. and Dickinson, J. (1978), 'Obstacles for the Development of a Capitalist Agriculture', *Journal of Peasant Studies*. 5:4, 446–481.

Margulis, L. and Sagan, D. (2002), *Acquiring Genomes: A Theory of the Origins of Species* (New York: Perseus Books Group).

Marks and Spencer 2007, 'This is Not Just Good Food – This is M&S Food', <http:// www.marksandspencer.com/gp/browse.html/ref=sc_ca_c_4_47368031_3/026-1436895-9226857?ie=UTF8&node=46484031&no=47368031&mnSBrand=cor e&me=A2BO0OYVBKIQJM>, accessed 16 May 2007.

Marsden, T.K. (2003), *The Condition of Rural Sustainability* (Van Goram, The Netherlands: European Perspectives in rural Development Series).

Marsden, T. and Smith, E. (2005), 'Ecological Entrepreneurship: Sustainable Development in Local Communities through Quality Food Production and Local Branding', *Geoforum* 36:4, 440–451.

—— (2007), 'Dealing with Complexity and Contingency in the Geographies of Agri-food: From Food Regimes to Sustainable Food Chains?' AAG Annual Conference Paper Presentation, San Francisco, April 17–21.

Marshall, E. (2000), 'Property Claims: A Deluge of Patents Creates Legal Hassles for Research', *Science* 288:5464, 255–257.

Massey, D. (1997), 'Power-Geometry and a Progressive Sense of Place', in Bird (eds).

—— (2006), 'Is the World Really Shrinking?', BBC interview, 5 November 2006.

Maxey, L. (2007), 'From "Alternative" to "Sustainable" Food', in Maye et al. (eds).

Maye, D., et al. (eds) (2007), *Constructing Alternative Food Geographies: Representation and Practice* (London: Elsevier Press).

McAfee, K. (2003), 'Neoliberalism on the Molecular Scale. Economic and Genetic Reductionism in Biotechnology Battles' *Geoforum* 34, 203–219.

—— (2004) 'Geographies of Risk and Difference in Crop Genetic Engineering', *The Geographical Review* 94:1, 80–107.

McCarthy, J. (2002), 'First World Political Ecology: Lessons From the Wise Use Movement', *Environment and Planning A* 34:7, 1281–1302.

McGinnis, J. Michael, et al (2006), 'Food Marketing to Children and Youth: Threat or Opportunity?', Committee on Food Marketing and the Diets of Children and Youth. National Academies Press. <http://www.nap.edu/catalog/11514.html>, accessed 25 October 2007.

McGirt, E. (2006), 'A Banner Year', *Fortune 500* 4 April 2006 <http://money.cnn.com/2006/03/31/news/companies/intro_f500_fortune/index.htm>, accessed 24 October 2007.

McKibben, B. (2003), *Enough: Staying Human in an Engineered Age* (New York: Times Books).

McLeod, D. (1976), 'Urban-Rural Food Alliances: A Perspective on Recent Community Food Organizing', Merrill, R. (ed.).

McNeal, J. (1992), *Kids as Customers: A Handbook of Marketing to Children* (New York: Lexington Books).

Merchant, C. (1996), 'Reinventing Eden: Western Culture as a Recovery Narrative', in Cronon, (ed.).

Merrill, R. (ed.) (1976), *Radical Agriculture* (New York: Harper Row).

Merz, J. (2000), 'Statement to the Subcommittee on Courts and Intellectual Property of the Committee on the Judiciary US House of Representatives: Oversight Hearings on Gene Patents and Other Genomic Inventions' <http://www.bioethics.upenn.edu/prog/ethicsgenes/>, accessed 24 October 2007.

Messer K. et al. (2006), 'Advertising Could Calm Food Safety Fears', Food USA Navigator.com, <http://www.foodnavigator-usa.com/news/ng.asp?n=69841&m=1FNU814&c=orpfwwsahpchdef>, accessed 1 May 2007.

Meter, K. (2006), 'Local Economic Analysis: Stronger Regions, Better Food Policies', Crossroads Resource Center (Minneapolis), paper presented at the Community

Food Security Coalition Conference, Vancouver, BC, 9 October 2006, <http://www.crcworks.org/crcppts/cfsc06.pdf>, accessed 2 February 2007.

Miller, M. et al (2006), 'Organic Agriculture in Wisconsin: 2005 Status Report', <http://www.cias.wisc.edu/pdf/organic05.pdf>, accessed 19 September 2007.

Ministry of Health (1924), 'Report of the Departmental Committee on the Use of Preservatives and Colouring Matters in Food' (London: HMSO).

Minou, Y. and Willer, H. (2003), *The World of Organic Agriculture 2003 – Statistics and Future Prospects*, <http://www.ifoam.org/>, accessed 24 October 2007.

Monsanto (2007), 'Monsanto Imagine: Roundup Ready YieldGard Plus With Roundup Ready Corn 2', <http://www.monsanto.ca/monsanto/layout/seedsandtraits/rr_yieldgard_plus_rrc2.asp>, accessed 31 May 2007.

Monforte Dairy (2007), Monforte Dairy (home page), < http://www.monfortedairy.com/.

Morgan, K. and Murdoch, J. (2000), 'Organic Versus Conventional Agriculture: Knowledge, Power and Innovation in the Food Chain', *Geoforum* 31:2, 159–173.

—— and Morley A. (2001), *Relocalising the Food Chain: The Role of Creative Public Procurement* (Cardiff: The Regeneration Institute, Cardiff University).

—— (2003), *School Meals: Healthy Eating & Sustainable Food Chains.* (The Regeneration Institute: Cardiff University).

—— et al. (2006), *Worlds of Food: Place, Power and Provenance in the Food Chain.* (Oxford: Oxford University Press).

—— (2006), 'The Forgotten Foodscape: School Canteens, Social Justice and Sustainable Development', paper presented at the *Annual Association of American Geographers Conference*, Chicago, IL, March 11, 2006.

Murdoch, J. and Miele, M. (1999), 'Back to Nature': Changing "Worlds of Production" in the Food Sector', *Sociologia Ruralis* 39:4, 465–483.

—— et al. (2000), 'Quality, Nature, and Embeddedness: Some Theoretical Considerations in the Context of the Food Sector', *Economic Geography* 76:2, 107–125.

—— (2006), *Post-structuralist Geography* (London: Sage Publications).

Muzart, G. (1999), 'Description of National Innovation Surveys Carried Out, or Foreseen, in 1997–99 in OECD Non-CIS-2 Participants and NESTI Observer Countries', *OECD Science, Technology and Industry Working Papers* (OECD Publishing).

National Biotechnology Advisory Committee (NBAC) (1999), *National Biotechnology Advisory Committee: Sixth Report, 1998* (Ottawa: Industry Canada).

National Cancer Institute (2007), *Milestone 1971: National Cancer Act*, <http://dtp.nci.nih.gov/timeline/noflash/milestones/M4_Nixon.htm>, accessed 29 March 2007.

National Farmers Union (NFU) (2006), 'Submission by the National Farmers Union Region 6 (Saskatchewan) on Selected Rural and Agricultural Issues to the Government of Saskatchewan', <http://www.nfu.ca/briefs.html#briefs>, accessed 9 February 2007.

—— (2005a), 'Policy on Sustainable Agriculture' <http://www.nfu.ca/sustag.htm>, accessed 9 February 2007.

—— (2005b), 'The Farm Crisis: Its Causes and Solutions', <http://www.nfu.ca/briefs.html#briefs>, accessed 9 February 2007.

—— (2005c), 'Solving the Farm Crisis: A Sixteen-Point Plan for Canadian Farm and Food Security', <http://www.nfu.ca/briefs.html#briefs>, accessed 9 February 2007.

National Institute on Media and the Family (2005), Mediawise Story – Annual Report. <http://www.mediafamily.org/about/annual_report_2005.pdf>, accessed 27 October 2007.

Negin, E. (1996), 'The Alar "Scare" was for Real', *Columbia Journalism Review*, September/October, <http://backissues.cjrarchives.org/year/96/5/alar.asp>, accessed 18 June 2007.

Nelson, R. (2004), 'The Market Economy, and the Scientific Commons', *Research Policy* 33:3, 455–471.

Nestle, M. (2006), 'Food Marketing and Childhood Obesity – A Matter of Policy', *New England Journal of Medicine* June 354, 2527–2529.

—— (2003), *Food Politics: How the Food Industry Influences Nutrition and Health*, (Berkeley: University of California Press).

Nierenberg, D. (2007), 'Preventing Avian Flu: Bigger Isn't Always Better', July 18, <http://www.worldwatch.org/node/5198>, accessed 1 August 2007.

Norris, M. (2006). 'Even the Army is Going Organic: Commissaries Go Organic', *Organic Consumers Association*, 10 March 2006. <www.organicconsumers.org/articles>, accessed 7 March 2007.

O'Brien, T. (1997), 'Factory Farming and Human Health', Report for Compassion in World Farming Trust. < http://www.ciwf.org.uk/>, accessed 3 June 2007.

Ofcom (2004), 'Childhood Obesity in Food Advertising in Context Children's food choices, parents' understanding and influence, and the role of food promotion', <http://www.ofcom.org.uk/research/tv/reports/food_ads/report.pdf>, accessed 15 May 2007.

Office International des epizooties – World Organization for Animal Health (OIE) (2006), 'Terrestrial Animal Health Code – 2006', <http://www.oie.int/eng/normes/mcode/en_chapitre_2.3.13.htm>, accessed 12 July 2007.

—— (2006) "Geographical Distribution of Countries that reported BSE Confirmed Cases since 1989". <http://www.oie.int/eng/info/en_esbcarte.htm>, BSE incidence by country: <http://www.oie.int/eng/info/en_esb.htm>, accessed 24 October 2007.

Oliver, D. and Fairbairn, J. (2004), 'Standing Committee on Agriculture and Forestry: Interim Report', The BSE Crisis – Lessons for the Future. <http://www.parl.gc.ca/37/3/parlbus/commbus/senate/com-e/agri-e/rep-e/repintapr04-e.pdf>, accessed 6 July 2007.

Ontario Federation of Agriculture (OFA) (2004), 'The Public Interest – Growing Ontario's Greenest Industry: Agriculture in Perspective', <http://www.ofa.on.ca/whatwedo/lobby/recentBriefs/prebudget submission to Minister of Finance.pdf>, accessed 11 March 2007.

—— (2006a), 'Farmers Feed Cities', <http://www.farmersfeedcities.com/>, accessed 12 February 2007.

Pavitt, K. (1991), 'What Makes Basic Research Economically Useful?' *Research Policy* 20, 109–119.

Pawlick, T. (2006), *The End of Food: How the Food Industry is Destroying Our Food Supply* (Toronto: Greystone Press).

Pearce, H. et al. (2005), 'Double Dividend? Promoting Good Nutrition and Sustainable Consumption Through School Meals', Commissioned by the Sustainable Consumption Roundtable and written by the Soil Association. A joint initiative from SDC and NCC. <http://www.sd-commission.org.uk/publications/downloads/Double_Dividend.pdf>, 25 September 2007.

Peeples, M. (2007), 'States Ban Catfish Imports From China Over Tests: All Things Considered', National Public Radio, Food, 11 May 2007, <http://www.npr.org/templates/story/story.php?storyId=10141686>, accessed 24 May 2007.

Pellerin, M. (2006), 'Politiquement alimentaire', Plenary Session IV: New Realities: What Are the Implications of Canadian Consumer Food Choices?: The McGill Institute for the Study of Canada Annual Conference: What Are We Eating? Towards a Canadian Food Policy, Montreal, February 15–17.

Penker, M. (2006), 'Mapping and Measuring the Ecological Embeddedness of Food Supply Chains', *Geoforum* 37, 368–379.

Perkel, C. (2007), 'Ontario Farmer, Raided for Selling Raw Milk, Pledges to Stay on Hunger Strike', published 28 November 2007 in the Canadian Press <http://www.cbc.ca/cp/national/061128/n112855A.html>, accessed 21 September 2007.

Phillips, P. (2002), 'Biotechnology in the Global Agri-food System', *Trends in Biotechnology* 20:9, 376–381.

Polanyi, K. (2001), *The Great Transformation: The Political and Economic Origins of Our Time* (Boston, Beacon Press).

Pollan, M. (2006), *Organics goes Mainstream.* Canadian Broadcasting Corporation (CBC) Ideas Series, Produced by Jill Eisen. <http://www.cbc.ca/ideas/features/organics/index.html>, accessed 7 December 2006.

Post, D. (2006), 'The Precautionary Principle and Risk Assessment in International Food Safety: How the World Trade Organization Influences Standards' *Risk Analysis* 26:5, 1259–1273.

Potter, M. (ed.) (2006), *Food Consumption and Disease Risk – Consumer – Pathogen Interactions* (Cambridge, England: Woodhead Publishing).

Poultry Industry Council of Canada (2005), 'No Need for Bird Flu to Become Bird Fear – Ontario Agriculture's Perspective on Avian Influenza', <http://www.poultryindustrycouncil.ca/avian_influenza.html>, accessed 28 May 2007.

Powell, H. (1956), *The Original Has This Signature – WK Kellogg* (Englewood Cliffs, NJ: Prentice Hall).

Power, E. (2005), 'The Determinants of Healthy Eating Among Low-income Canadians'. *Canadian Journal of Public Health 96:S3, S37-S42. www.hc-sc.gc.ca/hpfb-dgpsa/onpp-bppn/research_healthy_eating_e.html#syn*

Pretty, J. et al. (2005), 'Farm Costs and Food Miles: An Assessment of the Full Cost of the UK Weekly Food Basket', *Food Policy* 30:1, 1–19.

Priest. S. et al. (2003), 'The "Trust Gap" Hypothesis: Predicting Support for Biotechnology Across National Cultures as a Function of Trust in Actors', *Risk Analysis* 23:4, 751–766.

Raynolds, L. (2004), 'The Globalization of Organic Agro-Food Networks', *World Development* 32:5, 725–743.

Rees, G. (1969), *St Michael: A History of Marks and Spencer* (London: Weidenfeld and Nicolson).

Reid, L. (2003), 'History of Corn Breeding on the Central Experimental Farm, Ottawa', Agriculture and Agri-food Canada Central Experimental Farm, Ottawa <http://sci.agr.ca/ecorc/zea/zea03_e.htm>, accessed 25 October 2007.

Rey, P. and Winter, W. (1998), 'The Law and Economics of Tying Arrangements: Lessons for the Competition Policy Treatment of Intellectual Property', in Anderson and Gallini (eds).

Reynolds, C. (2004), 'Frontenac's Feast of Fields Promotes Local Growers', Eastern Agri-News, <http://www.agrinewsinteractive.com/archives/article-6056.htm>, accessed 11 Mar. 2007.

Richardson, P. (2004), 'Agricultural Ethics, Neurotic Natures and Emotional Encounters: An Application of Actor-network Theory', *Ethics Place and Environment* 7:3, 195–202.

Rifkin, J. (1981), *Algeny* (New York: The Viking Press).

Rimal, A. et al. (2005), 'Agro-biotechnology and Organic Food Purchase in the United Kingdom', *British Food Journal* 107:2, 84–97.

Ritzer, G. (2006), *McDonaldization: The Reader*, 2nd edition (Thousand Oaks: Pine Forge Press).

Roberts, W. (2006), 'Encounter in Food Security Conference', Panel Speaker on Urban Food Security, Ryerson University, June 22–24.

Robbins, P. (2004), *Political Ecology* (London: Blackwell Press).

Robin, C. (2004), 'The Politics and Antipolitics of Fear', *Raritan* 23:4, 79–108.

Robinson, G. (2006), 'Ontario's Environmental Farm Plan: Evaluation and Research Agenda', *Geoforum* 37:5, 859–873.

Rodgers, K.E. (1996), 'Multiple Meanings of Alar After the Scare: Implications for Closure', *Science, Technology & Human Values* 21, 177–197.

Rothstein, H. (2004), 'Precautionary Bans or Sacrificial Lambs? Participative Risk Regulation and the Reform of the UK Food Safety Regime', *Public Administration* 82:4, 857–881.

Rose, S. (2007), 'Back in Fashion: How We're Reviving a British Icon', *Harvard Business Review* 81:5, 51–58.

Rothschild, L. (1971), *A Framework for Government Research and Development* Cmnd. 4814.

Rousa, T. and Hunt, A. (2004), 'Governing Peanuts: The Regulation of the Social Bodies of Children and the Risks of Food Allergies', *Social Science & Medicine* 58, 825–836.

Royal Society of Canada (2001), 'Elements of Precaution: Recommendations for the Regulation of Food Biotechnology in Canada', An Expert Panel Report on the Future of Food Biotechnology, Report prepared for Health Canada, Canadian Food Inspection Agency and Environment Canada.

Rushkoff, Douglas (1999), *Coercion: Why We Listen to What "They" Say* (New York: Penguin Putnam).

Saskatchewan Organic Directorate (2006), 'Position Paper on the Introduction of Genetically Modified Alfalfa', http://www.saskorganic.com/oapf/pdf/SOD_GMO_Alfalfa_Position_Paper.pdf, accessed 25 May 2007.

—— (2007a), 'Organic Agriculture Protection Fund', <http://www.saskorganic.com/oapf/#latest>, accessed 25 May 2007.

—— (2007b), 'Position Paper on Seed Variety Registration in Canada', Prepared as input for public consultations on the new regulation being proposed by the Canadian Food Inspection Agency February 2007, <http://www.saskorganic.com/oapf/pdf/SOD-Position-Paper-Seed-Variety.pdf>, accessed 25 May 2007.

Sassatelli, R. (2006), 'Empowering Consumers: The Creative Procurement of School Meals in Italy and the UK'. <http://www.matforsk.no/web/wakt.nsf/ee079377855e855ec1256dbb002f1e6b/e5c31b2792458b3ec12571640033ffbe/$FILE/Empowering%20Consumers.doc>, accessed 27 August 2006.

Scapp, R. and Seitz, B. (1998), 'Introduction', in Scapp and Seitz (eds).

—— (eds) (1998), *Eating Culture* (Albany, New York: State University of New York Press).

Schlosser, E. (2002), *Fast Food Nation: The Dark Side of the All American Meal* (New York: Perennial Press).

Scotchmer, S. (1991), 'Standing on the Shoulders of Giants: Cumulative Research and the Patent Law', *Journal of Economic Perspectives* 5:1, 29–41.

—— (1998), 'R&D Joint Ventures and Other Cooperative Arrangements', Anderson and Gallini (eds).

Scott, A. et al. (2001), 'The Economic Returns to Basic Research and the Benefits of University-Industry Relationships: A Literature Review and Update of Findings', *Report for the Office of Science and Technology*. Science and Technology Policy Research, University of Sussex.

Scott, V. and Elliot, P. (2006), 'Influence of Food Processing Practices and Technologies on Consumer-pathogen Interactions', in Potter (ed.).

Severson, K. (2007), 'Lunch With Alice Waters, Food Revolutionary', *New York Times*, September 19, http://www.nytimes.com/2007/09/19/dining/19wate.html, accessed 29 October 2007.

Shilling, C. (2003), *Body and Social Theory* (London: Sage Publications).

Shughart II, W. (ed.) (1997), *Taxing Choice* (Northridge: California State University).

Sinclair, U. (1906, reprinted 2003) *The Jungle* (New York: Doubleday).

Slow Food Canada (2007), 'Contacts', <http://slowfood.ca/contact.php>, accessed 25 September 2007.

Slow Food USA. (2007), 'Home' <http://www.slowfoodusa.org/about/index.html>, accessed 25 September 2007.

Smith, A. (ed.) (2004), *The Oxford Encyclopaedia of Food and Drink in America* (New York: Oxford University Press).

—— and Watkiss, P. (2005), 'The Validity of Food Miles as an Indicator of Sustainable Development', report prepared for the *Department of the Environment, Food and*

Rural Affairs, UK. <http://statistics.defra.gov.uk/esg/reports/foodmiles/final. pdf>, accessed 24 October 2007.

Smith, M. (1991), 'From Policy Community to Issue Network: Salmonella in Eggs and the New Politics of Food', *Public Administration* 69:2, 235–255.

Smithers, J. and Blay-Palmer, A. (2001), 'Technology Innovation as a Strategy for Climate Adaptation in Agriculture', *Applied Geography* 21:2, 175–197.

Smithers, J. and Furman, M. (2003), 'Environmental Farm Planning in Ontario: Exploring Participation and the Endurance of Change', Land Use Policy 20, 343–356.

—— et al. (2005), 'Across the Divide (?): Reconciling Farm and Town Views of Agriculture-Community Linkages', *Journal of Rural Studies* 21:3, 281–295.

de Somogyi, J. (1967), '"St. Michael" – A Marketing Philosophy', *European Journal of Marketing*, 1:2, 54–59.

Sonnino, R. (2006), 'Embeddedness in Action: Saffron and the Making of the Local in Southern Tuscany', *Agriculture and Human Values* 24, 61–74.

—— and Marsden, T. (2006), 'Beyond the Divide: Rethinking the Relationships Between Alternative and Conventional Food Networks in Europe', *Journal of Economic Geography* 6:2, 181–199.

Stacey, M. (1994), *Consumed: Why Americans Love, Hate and Fear Food* (New York: Simon and Schuster).

Stamp Dawkins, M. et al. (2004), 'Chicken Welfare is Influenced More by Housing Conditions than by Stocking Density', *Nature* 427: 342–344.

State Library of Iowa. (2005), *Urban and Rural Population (1850–2000) and Metropolitan and Non-metropolitan Population (1950–2003)*. <http://data. iowadatacenter.org/browse/urbanruralareas.html#Population>, accessed 15 September 2006.

Statistics Canada (2001), 'Farms, by Farm Type and Province (2001 and 2006 Censuses of Agriculture)', (updated 16 May 2005), <http://www40.statcan.ca/ l01/cst01/agrc35a.htm, accessed 24 May 2007.

—— (2001a), 'Income of farm families', 2001 Census of Agriculture (updated 12 December 2003) <http://www.statcan.ca/english/agcensus2001/first/socio/ income.htm>, accessed 1 June 2007.

—— (2001b), 2001 Canadian Community Profiles: Kingston CMA. (updated 02 January 2007) <http://www12.statcan.ca/english/profil01/CP01/Details/Page. cfm?Lang=E&Geo1=CMA&Code1=521__&Geo2=PR&Code2=35&Data=Cou nt&SearchText=Kingston&SearchType=Begins&SearchPR=01&B1=All&Custo m>, accessed 30 September 2006.

—— (2004), 'Farm Census Family by Size, by Province (2001 Censuses of Agriculture and Population)', <http://www40.statcan.ca/l01/cst01/agrc41a.htm>, accessed 24 May 2007.

—— (2005), 'Canadian Statistics: Population Urban and Rural, by Province and Territory – Canada, (updated 1 September 2005) <http://www40.statcan. ca/l01/cst01/demo62a.htm?searchstrdisabled=1951&filename=demo62a. htm&lan=eng>, accessed 21 May 2006.

—— (2007), 'Net Farm Income 2006', *The Daily*, 28 May, (updated 28 May 2007) <http://www.statscan.ca/Daily/English/070528/d070528a.htm>, accessed 24 October 2007.

—— (2007a), 'Snapshot of Canadian Agriculture' (updated 26 October 2007) <http://www.statscan.ca/english/agcensus2006/articles/snapshot.htm>, accessed 24 May 2007.

—— (2007b), 'The Financial Picture of Farms in Canada', (updated 26 October 2007) <http://www.statscan.ca/english/agcensus2006/articles/finpic.htm>, accessed 24 May 2007.

Steckle, P. (2004), 'Canadian Live Stock and Beef Pricing in the Aftermath of the BSE Crisis', Report of the Standing Committee on Agriculture and Agri-Food, <http://cmte.parl.gc.ca/Content/HOC/committee/373/agri/reports/rp1282498/agrirp02/03-cov2-e.htm>, accessed 12 August 2007.

Story, M. and French, S. (2004), 'Food Advertising and Marketing Directed at Children and Adolescents in the US', *International Journal of Behavioural Nutrition and Physical Activity*, <http://www.ijbnpa.org/content/1/1/3>, accessed 21 May 2007.

Strasburger, V. and Wilson, B. (2002), *Youth and Media: Opportunities for Development or Lurking Dangers? Children, Adolescents and the Media* (Thousand Oaks, CA: Sage Publications).

Sumner, J. (2005), *Sustainability and the Civil Commons: Rural Communities in the Age of Globalization* (Toronto: University of Toronto Press).

Swyngedouw, E. (1997), 'Neither Global nor Local: "Glocalization" and the Politics of Scale', in Cox, (ed.).

Ten Eyck, T. (1999), 'Shaping a Food Safety Debate: Control Efforts of Newspaper Reporters and Sources in the Food Irradiation Controversy', *Science Communication* 20, 426–447.

Ten Eyck, T. and Deseran, F. (2001), 'In the Words of Experts: the Interpretive Process of the Food Irradiation Debate', *International Journal of Food Science and Technology* 36:8, 821–831.

Time Magazine (1941), 'Hunger: Cover Story', <http://www.time.com/time/magazine/article/0,9171,765774,00.html>, accessed 6 October 2007.

—— (1944) 'DDT' <http://www.time.com/time/magazine/article/0,9171,775033-1,00.html>, accessed 30 January 2007.

Toole, A. (2000), 'The Impact of Public Basic Research on Industrial Innovation: Evidence from the Pharmaceutical Industry', Stanford Institute for Economic Policy Research, SIEPR Discussion Paper No. 00-07.

Tour de CSA (2007), 'Bike the Barns', <http://www.macsac.org/bikethebarns/>, accessed 29 October 2007.

Trader Joe's (2007), 'How We Do Business', <http://www.traderjoes.com/how_we_do_biz.html>, accessed 15 May 2007.

Tse, K. (1985), *Marks & Spencer: Anatomy of Britain's Most Efficiently Managed Company* (Oxford: Pergamon Press).

Tuan, Y. (1979), *Landscapes of Fear* (New York: Pantheon Books).

United Nations Human Development (UNDP) (2004), Human Development Report 2004 Cultural Liberty in Today's Diverse World http://hdr.undp.org/reports/

global/2004/?CFID=9015943&CFTOKEN=db14679f0dbbe4af-ED459C72-1321-0B50-356C8D415F678E77&jsessionid=e6304f8f2f0362865171, accessed 28 October 2007.

US Army Soldier Systems Center – Natick, MA (2007), <http://www.natick.army.mil/>, accessed 30 April 2007.

USDA (2000), 'Food Irradiation: A Safe Measure', from May-June 1998 FDA Consumer magazine, <http://www.fda.gov/opacom/catalog/irradbro.html>, accessed 14 August 2007.

—— (2007), Food Distribution Programs. 'Department of Defense Fresh Fruit and Vegetable Program' <http://www.fns.usda.gov/fdd/programs/dod/DoD_ FreshFruitandVegetableProgram.pdf#xml=http://65.216.150.153/texis/search/ pdfhi.txt?query=Department+of+Defense+Fresh&pr=FNS&order=r&cq=&id=4 592c39c33b>, accessed 10 October 2007.

USDA-ERS (2002a), 'United States Department of Agriculture, Economic Research Services: Briefing Room, Farm Structure: Questions and Answers' (updated 22 November 2002) <http://www.ers.usda.gov/Briefing/FarmStructure/Questions/ aging.htm>, accessed 27 May 2007.

United States Department of Health and Human Services (2004), 'Expanded "Mad Cow" Safeguards Announced to Strengthen Existing Firewalls Against BSE Transmission', <http://www.hhs.gov/news/press/2004pres/20040126.html>, accessed 3 October 2007.

United Way (2007), 'True Stories: United Way of Greater Toronto is Proud to Support The Sunshine Garden, Now in its Third Year of Operation'. <http://www.uwgt. org/who_we_help/true_stories_garden.htm>, accessed 10 October 2007.

Vallat, B. (2007), 'Protecting the world from emerging diseases linked to globalisation', editorial from the Director General, World Organization for Animal Health, (updated 09 August 2007) <http://www.oie.int/eng/edito/en_lastedito. htm>, accessed 12 July 2007.

vanDonkersgoed, E. (2006), 'Corner Post #430, Signs of Renewal in Agriculture', *Farm & Countryside Commentary*, subscription list: corner.post@sympatico.ca.

Vogel, J. (1992), 'Obesity and its Relation to Physical Fitness in the U.S. Military', *Armed Forces and Society*, 18:4, 497–513.

Walker, R. (2004), *The Conquest of Bread: 150 Years of Agribusiness in California* (New York: The New Press).

Wal-Mart (2007), Wal-Mart (home page), <http://www.walmartstores.com/ GlobalWMStoresWeb/navigate.do?catg=316>, accessed 29 October 2007.

Ward, B. and Lewis, J. (2002), 'Plugging the Leaks: Making the Most of Every Pound that Enters your Local Economy', New Economics Foundation, <http:// www.neweconomics.org/gen/uploads/PTL%20handbook.pdf>, accessed 10 August 2006.

Warner, M. (2006), 'Wal-Mart Eyes Organic Foods' *New York Times*, 12 May 2006, <www.nytimes.com>, accessed 8 December 2006.

Watts, M. and Goodman, D. (1997), 'Agrarian Questions: Global Appetite, Local, Metabolism: Nature, Culture, and Industry in Fin-de-Siecle Agro-food Systems', in Goodman and Watts (eds).

—— et al. (2005), 'Making reconnections in Agro-food Geography: Alternative Systems of Provision', *Progress in Human Geography* 29:1, 22–40.

Webster, R. and Hulse, D. (2004), 'Microbial Adaptation and Change: Avian Influenza', *Scientific and Technical Review of the OIE*. 23:2, 453–465.

Weiss, L. et al. (2006), 'Free Trade in Mad Cows: How to Kill a Beef Industry' *Australian Journal of International Affairs* 60:3, 376–399.

Whatmore, S. and Thorne, L. (1997), 'Nourishing Networks: Alternative Geographies of Food' in Goodman and Watts (eds).

—— (2002), *Hybrid Geographies: Natures, Cultures, Spaces* (London: Sage Publications).

—— et al. (2003), 'Guest Editorial: What's Alternative about Alternative Food?', *Environment and Planning A* 35:3, 389–391.

Williams, R. (1973), 'Base and Superstructure in Marxist Cultural Theory'. *New Left Review* 82, 3–16.

Willer, H and Youssefi, M. (2007), *The World of Organic Agriculture: Statistics and Emerging Trends 2006, International Federation of Agricultural Organic Movement*. DE-53117 Bonn.

Winson, A. (1993), *The Intimate Commodity: Food and the Development of the Agro-Industrial Complex in Canada* (Guelph, Canada: Garamond Press).

Winter, M. (2003), 'Embeddedness, the New Food Economy and Defensive Localism', *Journal of Rural Studies* 19:1, 23–32.

Wiser Earth (2007), Wiser Earth (home page) <http://wiserearth.org/>, accessed 10 October 2007.

Wolf, E. (2001), *Pathways of Power: Building an Anthropology of the Modern World* (Berkeley: University of California Press).

Wolfe D. and Gertler, M. (2001), 'Globalization and Economic Restructuring in Ontario: From Industrial Heartland to Learning Region?', *European Planning Studies* 9:5, 575–592.

World Health Organization (WHO) (2007), 'Epidemic and Pandemic Alert Response (EPR): Cumulative Number of Confirmed Human Cases of Avian Influenza A/(H5N1) Reported to WHO'. <http://www.who.int/csr/disease/avian_influenza/country/cases_table_2007_10_02/en/index.html>, accessed 3 October 2007.

Wright, S. (1994), *Molecular Politics: Developing American and British Regulatory Policy for Genetic Engineering, 1972–1982* (Chicago: Chicago Press).

Wrigley, N. (2002), 'Food Deserts' in British Cities: Policy Context and Research Priorities', *Urban Studies* 39:11, 2029–2040.

—— and Currah, A. (2004). 'Networks of Organizational Learning and Adaptation in Retail TNCs', *Global Networks*, 4, 1–23.

—— et al. (2005), 'Globalizing Retail: Conceptualizing the Distribution-based TNC', *Progress in Human Geography*, 29, 437–457.

Yale (2006), *Organic Options Growing Across Yale*, <http://www.yale.edu/sustainability/foodproject.htm>, accessed 10 August 2006.

Young, J. (1981), 'The Long Struggle for the 1906 Law'. *US Food and Drug Administration: Department of Health and Human Services*. <http://vm.cfsan.fda.gov/~lrd/history2.html>, accessed 25 September 2007.

Index

Page numbers in *italics* refer to tables.